D0849858

Nutshell Series

of

WEST PUBLISHING COMPANY

P.O. Box 43526

St. Paul, Minnesota 55164

June, 1984

———

Accounting—Law and, 1984, 377 pages, by E. McGruder Faris, Professor of Law, Stetson University.

Administrative Law and Process, 2nd Ed., 1981, 445 pages, by Ernest Gellhorn, Dean and Professor of Law, Case Western Reserve University and Barry B. Boyer, Professor of Law, SUNY, Buffalo.

Admiralty, 1983, 390 pages, by Frank L. Maraist, Professor of Law, Louisiana State University.

Agency-Partnership, 1977, 364 pages, by Roscoe T. Steffen, Late Professor of Law, University of Chicago.

American Indian Law, 1981, 288 pages, by William C. Canby, Jr., Adjunct Professor of Law, Arizona State University.

Antitrust Law and Economics, 2nd Ed., 1981, 425 pages, by Ernest Gellhorn, Dean and Professor of Law, Case Western Reserve University.

Appellate Advocacy, 1984, 325 pages, by Alan D. Hornstein, Professor of Law, University of Maryland.

Art Law, 1984, 335 pages, by Leonard D. DuBoff, Professor of Law, Lewis and Clark College, Northwestern School of Law.

Banking and Financial Institutions, 1984, 409 pages, by William A. Lovett, Professor of Law, Tulane University.

Church-State Relations—Law of, 1981, 305 pages, by Leonard F. Manning, Late Professor of Law, Fordham University.

I

Civil Procedure, 1979, 271 pages, by Mary Kay Kane, Professor of Law, University of California, Hastings College of the Law.

Civil Rights, 1978, 279 pages, by Norman Vieira, Professor of Law, Southern Illinois University.

Commercial Paper, 3rd Ed., 1982, 404 pages, by Charles M. Weber, Professor of Business Law, University of Arizona and Richard E. Speidel, Professor of Law, Northwestern University.

Community Property, 1982, 423 pages, by Robert L. Mennell, Professor of Law, Hamline University.

Comparative Legal Traditions, 1982, 402 pages, by Mary Ann Glendon, Professor of Law, Boston College, Michael Wallace Gordon, Professor of Law, University of Florida and Christopher Osakwe, Professor of Law, Tulane University.

Conflicts, 1982, 469 pages, by David D. Siegel, Professor of Law, Albany Law School, Union University.

Constitutional Analysis, 1979, 388 pages, by Jerre S. Williams, Professor of Law Emeritus, University of Texas.

Constitutional Power—Federal and State, 1974, 411 pages, by David E. Engdahl, Professor of Law, University of Puget Sound.

Consumer Law, 2nd Ed., 1981, 418 pages, by David G. Epstein, Professor of Law, University of Texas and Steve H. Nickles, Professor of Law, University of Minnesota.

Contract Remedies, 1981, 323 pages, by Jane M. Friedman, Professor of Law, Wayne State University.

Contracts, 2nd Ed., 1983, 425 pages, by Gordon D. Schaber, Dean and Professor of Law, McGeorge School of Law and Claude D. Rohwer, Professor of Law, McGeorge School of Law.

Corporations—Law of, 1980, 379 pages, by Robert W. Hamilton, Professor of Law, University of Texas.

Corrections and Prisoners' Rights—Law of, 2nd Ed., 1983, 384 pages, by Sheldon Krantz, Dean and Professor of Law, University of San Diego.

Criminal Law, 1975, 302 pages, by Arnold H. Loewy, Professor of Law, University of North Carolina.

Criminal Procedure—Constitutional Limitations, 3rd Ed., 1980, 438 pages, by Jerold H. Israel, Professor of Law, University of Michigan and Wayne R. LaFave, Professor of Law, University of Illinois.

Debtor-Creditor Law, 2nd Ed., 1980, 324 pages, by David G. Epstein, Professor of Law, University of Texas.

Employment Discrimination—Federal Law of, 2nd Ed., 1981, 402 pages, by Mack A. Player, Professor of Law, University of Georgia.

Energy Law, 1981, 338 pages, by Joseph P. Tomain, Professor of Law, University of Cincinnatti.

Environmental Law, 1983, 343 pages by Roger W. Findley, Professor of Law, University of Illinois and Daniel A. Farber, Professor of Law, University of Minnesota.

Estate Planning—Introduction to, 3rd Ed., 1983, 370 pages, by Robert J. Lynn, Professor of Law, Ohio State University.

Evidence, Federal Rules of, 1981, 428 pages, by Michael H. Graham, Professor of Law, University of Illinois.

Evidence, State and Federal Rules, 2nd Ed., 1981, 514 pages, by Paul F. Rothstein, Professor of Law, Georgetown University.

Family Law, 1977, 400 pages, by Harry D. Krause, Professor of Law, University of Illinois.

Federal Estate and Gift Taxation, 3rd Ed., 1983, 509 pages, by John K. McNulty, Professor of Law, University of California, Berkeley.

Federal Income Taxation of Individuals, 3rd Ed., 1983, 487 pages, by John K. McNulty, Professor of Law, University of California, Berkeley.

Federal Income Taxation of Corporations and Stockholders, 2nd Ed., 1981, 362 pages, by Jonathan Sobeloff, Late Professor of Law, Georgetown University and Peter P. Weidenbruch, Jr., Professor of Law, Georgetown University.

Federal Jurisdiction, 2nd Ed., 1981, 258 pages, by David P. Currie, Professor of Law, University of Chicago.

Future Interests, 1981, 361 pages, by Lawrence W. Waggoner, Professor of Law, University of Michigan.

Government Contracts, 1979, 423 pages, by W. Noel Keyes, Professor of Law, Pepperdine University.

Historical Introduction to Anglo-American Law, 2nd Ed., 1973, 280 pages, by Frederick G. Kempin, Jr., Professor of Business Law, Wharton School of Finance and Commerce, University of Pennsylvania.

Immigration Law and Procedure, 1984, 345 pages, by David Weissbrodt, Professor of Law, University of Minnesota.

Injunctions, 1974, 264 pages, by John F. Dobbyn, Professor of Law, Villanova University.

Insurance Law, 1981, 281 pages, by John F. Dobbyn, Professor of Law, Villanova University.

Intellectual Property—Patents, Trademarks and Copyright, 1983, 428 pages, by Arthur R. Miller, Professor of Law, Harvard University, and Michael H. Davis, Professor of Law, Cleveland State University, Cleveland-Marshall College of Law.

International Business Transactions, 2nd Ed., 1984, 476 pages, by Donald T. Wilson, Professor of Law, Loyola University, Los Angeles.

Introduction to the Study and Practice of Law, 1983, 418 pages, by Kenney F. Hegland, Professor of Law, University of Arizona.

Judicial Process, 1980, 292 pages, by William L. Reynolds, Professor of Law, University of Maryland.

Jurisdiction, 4th Ed., 1980, 232 pages, by Albert A. Ehrenzweig, Late Professor of Law, University of California, Berkeley, David W. Louisell, Late Professor of Law, University of California, Berkeley and Geoffrey C. Hazard, Jr., Professor of Law, Yale Law School.

Juvenile Courts, 3rd Ed., 1984, 291 pages, by Sanford J. Fox, Professor of Law, Boston College.

Labor Arbitration Law and Practice, 1979, 358 pages, by Dennis R. Nolan, Professor of Law, University of South Carolina.

Labor Law, 1979, 403 pages, by Douglas L. Leslie, Professor of Law, University of Virginia.

Land Use, 1978, 316 pages, by Robert R. Wright, Professor of Law, University of Arkansas, Little Rock and Susan Webber, Professor of Law, University of Arkansas, Little Rock.

Landlord and Tenant Law, 1979, 319 pages, by David S. Hill, Professor of Law, University of Colorado.

Law Study and Law Examinations—Introduction to, 1971, 389 pages, by Stanley V. Kinyon, Late Professor of Law, University of Minnesota.

Legal Interviewing and Counseling, 1976, 353 pages, by Thomas L. Shaffer, Professor of Law, Washington and Lee University.

Legal Research, 4th Ed., 1984, approximately 425 pages, by Morris L. Cohen, Professor of Law and Law Librarian, Yale University.

Legal Writing, 1982, 294 pages, by Dr. Lynn B. Squires and Marjorie Dick Rombauer, Professor of Law, University of Washington.

Legislative Law and Process, 1975, 279 pages, by Jack Davies, Professor of Law, William Mitchell College of Law.

Local Government Law, 2nd Ed., 1983, 404 pages, by David J. McCarthy, Jr., Professor of Law, Georgetown University.

Mass Communications Law, 2nd Ed., 1983, 473 pages, by Harvey L. Zuckman, Professor of Law, Catholic University and Martin J. Gaynes, Lecturer in Law, Temple University.

Medical Malpractice—The Law of, 1977, 340 pages, by Joseph H. King, Professor of Law, University of Tennessee.

Military Law, 1980, 378 pages, by Charles A. Shanor, Professor of Law, Emory University and Timothy P. Terrell, Professor of Law, Emory University.

Oil and Gas, 1983, 443 pages, by John S. Lowe, Professor of Law, University of Tulsa.

Personal Property, 1983, 322 pages, by Barlow Burke, Jr., Professor of Law, American University.

Post-Conviction Remedies, 1978, 360 pages, by Robert Popper, Professor of Law, University of Missouri, Kansas City.

Presidential Power, 1977, 328 pages, by Arthur Selwyn Miller, Professor of Law Emeritus, George Washington University.

Procedure Before Trial, 1972, 258 pages, by Delmar Karlen, Professor of Law Emeritus, New York University.

Products Liability, 2nd Ed., 1981, 341 pages, by Dix W. Noel, Late Professor of Law, University of Tennessee and Jerry J. Phillips, Professor of Law, University of Tennessee.

Professional Responsibility, 1980, 399 pages, by Robert H. Aronson, Professor of Law, University of Washington, and Donald T. Weckstein, Professor of Law, University of San Diego.

Real Estate Finance, 2nd Ed., 1985, approximately 300 pages, by Jon W. Bruce, Professor of Law, Vanderbilt University.

Real Property, 2nd Ed., 1981, 448 pages, by Roger H. Bernhardt, Professor of Law, Golden Gate University.

Regulated Industries, 1982, 394 pages, by Ernest Gellhorn, Dean and Professor of Law, Case Western Reserve University, and Richard J. Pierce, Professor of Law, Tulane University.

Remedies, 2nd Ed., 1984, approximately 325 pages, by John F. O'Connell, Professor of Law, Western State University College of Law, Fullerton.

Res Judicata, 1976, 310 pages, by Robert C. Casad, Professor of Law, University of Kansas.

Sales, 2nd Ed., 1981, 370 pages, by John M. Stockton, Professor of Business Law, Wharton School of Finance and Commerce, University of Pennsylvania.

Schools, Students and Teachers—Law of, 1984, 409 pages, by Kern Alexander, Professor of Education, University of Florida and M. David Alexander, Professor, Virginia Tech University.

Sea—Law of, 1984, approximately 250 pages, by Louis B. Sohn, Professor of Law, Harvard University and Kristen Gustafson.

Secured Transactions, 2nd Ed., 1981, 391 pages, by Henry J. Bailey, Professor of Law Emeritus, Willamette University.

Securities Regulation, 2nd Ed., 1982, 322 pages, by David L. Ratner, Dean and Professor of Law, University of San Francisco.

Sex Discrimination, 1982, 399 pages, by Claire Sherman Thomas, Lecturer, University of Washington, Women's Studies Department.

Torts—Injuries to Persons and Property, 1977, 434 pages, by Edward J. Kionka, Professor of Law, Southern Illinois University.

Torts—Injuries to Family, Social and Trade Relations, 1979, 358 pages, by Wex S. Malone, Professor of Law Emeritus, Louisiana State University.

Trial Advocacy, 1979, 402 pages, by Paul B. Bergman, Adjunct Professor of Law, University of California, Los Angeles.

Trial and Practice Skills, 1978, 346 pages, by Kenney F. Hegland, Professor of Law, University of Arizona.

Trial, The First—Where Do I Sit? What Do I Say?, 1982, 396 pages, by Steven H. Goldberg, Professor of Law, University of Minnesota.

Unfair Trade Practices, 1982, 444 pages, by Charles R. McManis, Professor of Law, Washington University, St. Louis.

Uniform Commercial Code, 2nd Ed., 1984, approximately 500 pages, by Bradford Stone, Professor of Law, Detroit College of Law.

Uniform Probate Code, 1978, 425 pages, by Lawrence H. Averill, Jr., Dean and Professor of Law, University of Arkansas, Little Rock.

Water Law, 1984, 439 pages, by David H. Getches, Professor of Law, University of Colorado.

Welfare Law—Structure and Entitlement, 1979, 455 pages, by Arthur B. LaFrance, Dean and Professor of Law, Lewis and Clark College, Northwestern School of Law.

NUTSHELL SERIES

Wills and Trusts, 1979, 392 pages, by Robert L. Mennell, Professor of Law, Hamline University.

Workers' Compensation and Employee Protection Laws, 1984, 248 pages, by Jack B. Hood, Professor of Law, Cumberland School of Law, Samford University and Benjamin A. Hardy, Professor of Law, Cumberland School of Law, Samford University.

Hornbook Series

and

Basic Legal Texts

of

WEST PUBLISHING COMPANY

P.O. Box 43526

St. Paul, Minnesota 55164

June, 1984

Administrative Law, Davis' Text on, 3rd Ed., 1972, 617 pages, by Kenneth Culp Davis, Professor of Law, University of San Diego.

Agency and Partnership, Reuschlein & Gregory's Hornbook on the Law of, 1979 with 1981 Pocket Part, 625 pages, by Harold Gill Reuschlein, Professor of Law Emeritus, Villanova University and William A. Gregory, Professor of Law, Georgia State University.

Antitrust, Sullivan's Hornbook on the Law of, 1977, 886 pages, by Lawrence A. Sullivan, Professor of Law, University of California, Berkeley.

Common Law Pleading, Koffler and Reppy's Hornbook on, 1969, 663 pages, by Joseph H. Koffler, Professor of Law, New York Law School and Alison Reppy, Late Dean and Professor of Law, New York Law School.

Conflict of Laws, Scoles and Hay's Hornbook on, Student Ed., 1982, 1085 pages, by Eugene F. Scoles, Professor of Law, University of Illinois and Peter Hay, Dean and Professor of Law, University of Illinois.

Constitutional Law, Nowak, Rotunda and Young's Hornbook on, 2nd Ed., Student Ed., 1983, 1172 pages, by John E. Nowak, Professor of Law, University of Illinois, Ronald D. Rotunda, Professor of Law, University of Illinois, and J. Nelson Young, Professor of Law, University of North Carolina.

Contracts, Calamari and Perillo's Hornbook on, 2nd Ed., 1977, 878 pages, by John D. Calamari, Professor of Law, Fordham University and Joseph M. Perillo, Professor of Law, Fordham University.

Contracts, Corbin's One Volume Student Ed., 1952, 1224 pages, by Arthur L. Corbin, Late Professor of Law, Yale University.

Corporate Taxation, Kahn's Handbook on, 3rd Ed., Student Ed., Soft cover, 1981 with 1983 Supplement, 614 pages, by Douglas A. Kahn, Professor of Law, University of Michigan.

Corporations, Henn and Alexander's Hornbook on, 3rd Ed., Student Ed., 1983, 1371 pages, by Harry G. Henn, Professor of Law, Cornell University and John R. Alexander, Member, New York and Hawaii Bars.

Criminal Law, LaFave and Scott's Hornbook on, 1972, 763 pages, by Wayne R. LaFave, Professor of Law, University of Illinois, and Austin Scott, Jr., Late Professor of Law, University of Colorado.

Criminal Procedure, LaFave and Israel's Hornbook on, Student Ed., 1985, approximately 1300 pages, by Wayne R. LaFave, Professor of Law, University of Illinois and Jerold H. Israel, Professor of Law University of Michigan.

Damages, McCormick's Hornbook on, 1935, 811 pages, by Charles T. McCormick, Late Dean and Professor of Law, University of Texas.

Domestic Relations, Clark's Hornbook on, 1968, 754 pages, by Homer H. Clark, Jr., Professor of Law, University of Colorado.

Economics and Federal Antitrust Law, Hovenkamp's Hornbook on, Student Ed., 1985, approximately 375 pages, by Herbert Hovenkamp, Professor of Law, University of California, Hastings College of the Law.

Environmental Law, Rodgers' Hornbook on, 1977 with 1984 Pocket Part, 956 pages, by William H. Rodgers, Jr., Professor of Law, University of Washington.

Evidence, Lilly's Introduction to, 1978, 486 pages, by Graham C. Lilly, Professor of Law, University of Virginia.

Evidence, McCormick's Hornbook on, 3rd Ed., Student Ed., 1984, 1155 pages, General Editor, Edward W. Cleary, Professor of Law Emeritus, Arizona State University.

Federal Courts, Wright's Hornbook on, 4th Ed., Student Ed., 1983, 870 pages, by Charles Alan Wright, Professor of Law, University of Texas.

Federal Income Taxation of Individuals, Posin's Hornbook on, Student Ed., 1983, 491 pages, by Daniel Q. Posin, Jr., Professor of Law, Southern Methodist University.

Future Interest, Simes' Hornbook on, 2nd Ed., 1966, 355 pages, by Lewis M. Simes, Late Professor of Law, University of Michigan.

Insurance, Keeton's Basic Text on, 1971, 712 pages, by Robert E. Keeton, Professor of Law Emeritus, Harvard University.

Labor Law, Gorman's Basic Text on, 1976, 914 pages, by Robert A. Gorman, Professor of Law, University of Pennsylvania.

Law Problems, Ballentine's, 5th Ed., 1975, 767 pages, General Editor, William E. Burby, Late Professor of Law, University of Southern California.

Legal Writing Style, Weihofen's, 2nd Ed., 1980, 332 pages, by Henry Weihofen, Professor of Law Emeritus, University of New Mexico.

Local Government Law, Reynolds' Hornbook on, 1982, 860 pages, by Osborne M. Reynolds, Professor of Law, University of Oklahoma.

New York Practice, Siegel's Hornbook on, 1978, with 1981-82 Pocket Part, 1011 pages, by David D. Siegel, Professor of Law, Albany Law School of Union University.

Oil and Gas, Hemingway's Hornbook on, 2nd Ed., Student Ed., 1983, 543 pages, by Richard W. Hemingway, Professor of Law, University of Oklahoma.

Poor, Law of the, LaFrance, Schroeder, Bennett and Boyd's Hornbook on, 1973, 558 pages, by Arthur B. LaFrance, Dean and Professor of Law, Lewis and Clark College, Northwestern School of Law, Milton R. Schroeder, Professor of Law, Arizona State University, Robert W. Bennett, Professor of Law, Northwestern University and William E. Boyd, Professor of Law, University of Arizona.

Property, Boyer's Survey of, 3rd Ed., 1981, 766 pages, by Ralph E. Boyer, Professor of Law, University of Miami.

Property, Law of, Cunningham, Whitman and Stoebuck's Hornbook on, Student Ed., 1984, approximately 808 pages, by Roger A. Cunningham, Professor of Law, University of Michigan, Dale A. Whitman, Dean and Professor of Law, University of Missouri–Columbia and William B. Stoebuck, Professor of Law, University of Washington.

Real Estate Finance Law, Osborne, Nelson and Whitman's Hornbook on, (successor to Hornbook on Mortgages), 1979, 885 pages, by George E. Osborne, Late Professor of Law, Stanford University, Grant S. Nelson, Professor of Law, University of Missouri, Columbia and Dale A. Whitman, Dean and Professor of Law, University of Missouri, Columbia.

Real Property, Burby's Hornbook on, 3rd Ed., 1965, 490 pages, by William E. Burby, Late Professor of Law, University of Southern California.

Real Property, Moynihan's Introduction to, 1962, 254 pages, by Cornelius J. Moynihan, Professor of Law, Suffolk University.

Remedies, Dobb's Hornbook on, 1973, 1067 pages, by Dan B. Dobbs, Professor of Law, University of Arizona.

Sales, Nordstrom's Hornbook on, 1970, 600 pages, by Robert J. Nordstrom, former Professor of Law, Ohio State University.

Secured Transactions under the U.C.C., Henson's Hornbook on, 2nd Ed., 1979, with 1979 Pocket Part, 504 pages, by Ray D. Henson, Professor of Law, University of California, Hastings College of the Law.

Securities Regulation, Hazen's Hornbook on the Law of, Student Ed., 1985, approximately 550 pages, by Thomas Lee Hazen, Professor of Law, University of North Carolina.

Torts, Prosser and Keeton's Hornbook on, 5th Ed., Student Ed., 1984, 1286 pages, by William L. Prosser, Late Dean and Professor of Law, University of California, Berkeley, Page Keeton, Professor of Law Emeritus, University of Texas, Dan B. Dobbs, Professor of Law University of Arizona, Robert E. Keeton, Professor of Law Emeritus, Harvard University and David G. Owen, Professor of Law, University of South Carolina.

Trial Advocacy, Jeans' Handbook on, Student Ed., Soft cover, 1975, by James W. Jeans, Professor of Law, University of Missouri, Kansas City.

Trusts, Bogert's Hornbook on, 5th Ed., 1973, 726 pages, by George G. Bogert, Late Professor of Law, University of Chicago and George T. Bogert, Attorney, Chicago, Illinois.

Urban Planning and Land Development Control, Hagman's Hornbook on, 1971, 706 pages, by Donald G. Hagman, Late Professor of Law, University of California, Los Angeles.

Uniform Commercial Code, White and Summers' Hornbook on, 2nd Ed., 1980, 1250 pages, by James J. White, Professor of Law, University of Michigan and Robert S. Summers, Professor of Law, Cornell University.

Wills, Atkinson's Hornbook on, 2nd Ed., 1953, 975 pages, by Thomas E. Atkinson, Late Professor of Law, New York University.

Advisory Board

OIL AND GAS LAW
IN A NUTSHELL

By

JOHN S. LOWE

Professor of Law and Associate
Director of the National Energy
Law and Policy Institute,
The University of Tulsa.

ST. PAUL, MINN.
WEST PUBLISHING CO.
1983

COPYRIGHT © 1983 Jacquelyn Taft Lowe, custodian
50 West Kellogg Boulevard
P.O. Box 3526
St. Paul, Minnesota 55165

Library of Congress Cataloging in Publication Data

Lowe, John S., 1941–
  Oil and gas in a nutshell.

  (Nutshell series)
  Includes index.
  1. Petroleum law and legislation—United States.
2. Gas, Natural—Law and legislation—United States.
I. Title. II. Series.
KF1850.L68   1983   346.7304'6823   83–6811
                       347.30646823

ISBN 0–314–73469–4

To my parents,
John F. and Florence Lowe

*

PREFACE

The petroleum age did not begin until after the mid-nineteenth century. The first oil well in the United States was drilled in 1859 to a depth of 69 feet. In the approximately 125 years since then, the United States and the rest of the world have developed economies that are based on oil and gas as the major source of fuel. Approximately 43 percent of U.S. energy used comes from oil and another 26 percent from natural gas. On a worldwide basis, oil is even more important than in the United States. Natural gas is of lesser relative importance, but still a major source of energy and petrochemicals.

It is likely that the end of the petroleum age is in sight. We have probably discovered the easiest, the largest and the most productive of the world's deposits of oil and gas. Many say that we have used approximately half of the probable oil and gas resources of the United States, and that those remaining will be exhausted over the next forty to fifty years. Certainly, present reserves and likely resources cannot sustain world demand for more than a few generations.

However, it does not follow that oil and gas law is of waning importance. Those oil and gas resources which remain are likely to become even more valuable than they are today. A measure of this phenomenon was seen in the 1970's when curtailments of world supplies of oil and gas caused prices to increase tenfold in seven years. Because oil and gas are likely to increase in value, an understanding of the legal principles that control their development is likely to gain in importance.

This book focuses upon the legal rules that govern development of privately owned mineral rights. That is because most mineral rights in the United States are owned privately, though the federal and state governments own hundreds of millions of acres. Also, the rules for governmentally owned resources tend to be based on those for private transactions.

Because of the economic importance of oil and gas resources, there are numerous secondary sources and research materials available. Included in those that I have used in preparing this book are Kuntz, The Law of Oil and Gas (Anderson Publishing Co.); Williams and Meyers, Oil and Gas Law (Matthew Bender); Brown, The Law of Oil and Gas Leases (Matthew Bender); Hemingway, Oil and Gas (West); Sullivan, Handbook of Oil and Gas Law (Prentice-Hall); Summers, The Law of Oil and Gas (Vernon); and Merrill, Covenants Implied in Oil and Gas Leases, (Thomas Law Book Company). In addition, I have used many of the cases included in Williams, Maxwell, and Meyers, Cases on Oil and Gas (Foundation Press) and Huie, Woodward and Smith (West) as illustrations.

I want also to acknowledge the support of research assistants at The University of Tulsa who helped me research and prepare this book, including Dona K. Broyles, Gregory N. Fiske, Laurie A. Patterson, David Keehn, James Arlington, Kenneth L. Wire, David P. Page, Jeffrey R. Fiske, Arthur H. Adams, Tommy H. Butler, Michael F. Miller, Joseph G. Staskal, Curtis L. Craig, Michael D. Cooke, Steve E. McCain, Harley W. Thomas, Laura E. Frossard, Mark S. Rains and Thomas J. Wagner. Special thanks is due to Larry D. Vredenburgh, Ph.D., independent geologic consultant, who effectively rewrote my draft of Chapter 1, to Dr. Norman J. Hyne, Associate Professor of Geology at The University of Tulsa, who provided the diagrams for Chapter 1, to Dr. Joseph F. Fusco, Director of Instructional Media at The University of Tulsa, for his help with graphics, to James A. Hogue, Sr., Esq., for his comments on Chapter 13, and to Owen L. Anderson, Professor of Law at the University of North Dakota for his general review. Of course, defects in analysis or errors in statements are my own responsibility.

Finally, accolades are due to my wife, Jacquelyn, and my children, Sarah and Jack, for tolerating the domestic disruption caused by this undertaking, and to my secretary, Frances Kesely, for her diligent efforts in typing and retyping the many drafts of this manuscript.

JOHN S. LOWE

Tulsa, Oklahoma
May, 1983

*

OUTLINE

	Page
Preface -------------------------------	XIX
Table of Cases -------------------------	XLIII

**PART I. THE NATURE AND PROTEC-
TION OF OIL AND GAS
RIGHTS** -------------------- 1

Chapter 1. The Formation and Production of
Oil and Gas ----------------- 1
  A. Formation of Oil and Gas ---------- 1
  B. Drilling for Oil and Gas ----------- 3
  C. Producing Oil and Gas ------------- 5

Chapter 2. Ownership of Oil and Gas Rights 8
  A. The Ad Coelum Doctrine ----------- 8
  B. The Rule of Capture --------------- 9
  C. Limits to the Rule of Capture ------ 10
    1. Inherent Limitations ---------- 10
      a. Escaped Hydrocarbons ----- 10
      b. Drainage by Enhanced Re-
         covery Operations ------- 11
    2. Doctrine of Correlative Rights - 13
    3. Conservation Laws ------------ 14
      a. Economic and Physical
         Waste ----------------- 15
      b. Function of Oil and Gas
         Conservation Laws --- 17
      (1) Purpose -------------- 17
      (2) Well Spacing Rules ---- 18

Chapter 2.   Ownership of Oil and Gas Rights
—Continued   **Page**

(3) Production Regulation __   18
(4) Protection of Correlative
Rights _____   20

D.   Theories of Ownership of Oil and Gas   22
1.   Non-Ownership and Ownership
in Place Theories _____   22
2.   Significance of the Theories of
Ownership _____   24
a.   The   Corporeal/Incorporeal
Distinction _____   24
(1) Abandonment of Oil and
Gas Interests _____   24
(2) Forms of Action to Protect Oil and Gas Rights   27
b.   Classification of Oil and Gas
Rights as Real Property or
Personal Property _____   28
c.   Practical Impact of the Theory of Ownership _____   29

Chapter 3.   Kinds of Oil and Gas Interests __   31
A.   Fee Interest _____   31
B.   Mineral Interest _____   32
1.   Mineral Interest Includes Right
to Use the Surface _____   32
2.   Characteristics of Mineral Interest Ownership _____   33
3.   Louisiana's Mineral Servitude __   35
C.   Leasehold Interest/Working Interest   36
D.   Surface Interest _____   36

Chapter 3. Kinds of Oil and Gas Interests
—Continued   **Page**
   E. Royalty Interest _____ 37
     1. Kinds of Royalty Interests _____ 38
     2. Characteristics of Royalty Interests _____ 39
   F. Production Payment _____ 40
   G. Net Profits Interest _____ 40
   H. Carried Interest _____ 41
   I. Other Interests _____ 43

Chapter 4. Protection of Oil and Gas Rights 44
   A. Damage to Lease Value _____ 44
     1. Rationale of the Remedy _____ 45
     2. Measure of Damages _____ 46
   B. Slander of Title _____ 47
     1. False Claim _____ 47
     2. Malicious Intent _____ 47
     3. Specific Damages _____ 48
   C. Assumpsit _____ 48
   D. Conversion and Ejectment _____ 49
     1. Bad Faith Trespass _____ 49
     2. Good Faith Trespass _____ 50
   E. Conclusion _____ 51

PART II. CONVEYING OIL AND GAS RIGHTS _____ 52

Chapter 5. Creation and Transfer of Oil and Gas Interests _____ 52
   A. By Conveyance _____ 52
     1. Writing _____ 53
     a. Deeds _____ 53
     (1) Warranty Deed _____ 53

Chapter 5. Creation and Transfer of Oil and
Gas Interests—Con-
tinued   **Page**
(2) Quitclaim Deed 54
(3) Importance of Title Covenants 55
b. Oil and Gas Leases 56
c. Other Instruments 57
2. Words of Grant 57
3. Description 58
a. Legal Validity 58
b. Marketability 59
c. Methods of Description 59
(1) Reference to Government
Survey 59
(2) Metes and Bounds 60
4. Parties Designated 62
a. Identification of the Parties 62
b. Capacity of the Parties 63
5. Execution 63
a. Signature 63
b. Attestation and Acknowledgment 65
c. Delivery and Acceptance ... 67
d. Recording 69
B. By Inheritance 69
C. By Judicial Transfer 70
D. By Adverse Possession 70
1. Are Both the Surface and Minerals Adversely Possessed? 72
a. Unity of Possession 72
b. Relation Back 73
c. Paper Transactions 74

Chapter 5.   Creation and Transfer of Oil and
             Gas Interests—Continued   **Page**
        2.   What Must Be Done to Adverse-
             ly Possess Severed Minerals? _   74
        3.   Unresolved Issues _____   75
             a.   What It Takes to "Sever"
                  Minerals _____   76
             b.   How Much of the Mineral Is
                  Acquired _____   77
             c.   What Minerals Are Earned   78

Chapter 6.   Joint Ownership of Oil and Gas
             Rights _____   79
      A.   Concurrent Owners _____   79
        1.   Development   by   Concurrent
             Owners _____   80
             a.   Minority Rule _____   81
             b.   Majority Rule _____   81
        2.   A Critical Evaluation of the Ma-
             jority Rule _____   83
        3.   Other Methods of Obtaining the
             Right to Develop _____   85
             a.   Forced Pooling _____   85
             b.   Judicial Partition _____   86
             c.   Lost Mineral Interests _____   87
      B.   Marital Rights _____   90
      C.   Debtors/Creditors _____   92
      D.   Fiduciaries/Beneficiaries _____   94
      E.   Executive/Non-Executive Owners ___   95
      F.   Life Tenants/Remaindermen _____   96
        1.   Power to Grant _____   97
             a.   In Common Law States ____   97
             b.   In Louisiana _____   98

Chapter 6. Joint Ownership of Oil and Gas
Rights—Continued　**Page**
　　c.　Common Leasing Practice __　98
　2.　Division of Proceeds _____　100
　3.　The Open Mine Doctrine _____　101
G.　Term Interests _____　102

Chapter 7. Interpretive Problems in Oil and
Gas Conveyancing _____　105
A.　Steps in Judicial Interpretation _____　105
　1.　Interpretation of the Instrument
as a Whole _____　106
　2.　Use of Construction Aids or Circumstantial Tests _____　106
　3.　Consideration of Extrinsic Evidence _____　107
　4.　Application of the Interpretive
Steps _____　108
B.　What Is the Meaning of "Minerals" _　108
　1.　What the Courts Have Done ___　109
　2.　The Texas Experience _____　111
　　a.　*Acker v. Guinn*: The Surface
Destruction Test Articu-
lated _____　111
　　b.　*Reed v. Wylie I*: The Surface Destruction Test Re-
defined _____　112
　　c.　*Reed v. Wylie II*: Further
Refinement _____　113
　3.　A Proposed Solution _____　115
C.　The Mineral/Royalty Distinction ____　117
　1.　The Significance of the Distinction _____　117

Chapter 7. Interpretive Problems in Oil and
            Gas Conveyancing—Continued **Page**
         2. Common Interpretive Problems 118
             a. Guidelines to Interpretation 119
             b. Avoiding Ambiguity 122
     D. Fractional Interest Problems 123
         1. Double Fractions 123
             a. Guidelines to Interpretation 124
             b. Avoiding Ambiguity 126
         2. Overconveyance 126
             a. The *Duhig* Rule 127
             b. Departures From the Rule . 129
             c. Application to Leases as
                 Well as Deeds 131
             d. Avoiding the Overconveyance
                 Problem 132
         3. Mineral Acres/Royalty Acres .. 133
             a. Mineral Acres 133
             b. Royalty Acres 134
     E. Conveyances of Leased Property 134
         1. The "Subject To" Problem 135
             a. Purpose of the "Subject To"
                 Clause 135
             b. The Two Grants Doctrine .. 135
             c. Avoiding the "Subject To"
                 Ambiguity 137
         2. Apportionment of Royalties 138
             a. The Non-Apportionment Rule 140
             b. The Apportionment Rule ... 141
             c. Understanding the Rules ... 141

Chapter 7. Interpretive Problems in Oil and
          Gas Conveyancing—Con-
          tinued   **Page**

     d. Avoiding Conflict With the
         Rules 143
         (1) Modification by Agree-
             ment 143
         (2) Entirety Clauses 144
         (3) Legislative Provisions .. 146
     3. Top Leasing 147
     a. The Rule Against Perpetuities 148
     b. Obstruction 148

PART III.   OIL AND GAS LEASING 150

Chapter 8. Essential Clauses of Modern Oil
          and Gas Leases 150
  A. Purpose of the Lease 150
  B. The Nature of the Lease 152
     1. Both a Conveyance and a Con-
         tract 152
     2. More a Deed Than a Lease 153
     3. Legal Classification 153
     4. Essential Provisions 153
  C. Granting Clause 154
     1. Size of the Interest Granted ... 154
     2. Substances Covered by the Grant 155
     3. Land Covered by the Lease: The
         Mother Hubbard Problem ... 156
     4. Uses Permitted by the Grant .. 159
     a. General Principle: Lessee's
         Right to Use Burdens the
         Surface 159

Chapter 8. Essential Clauses of Modern Oil and Gas Leases—Continued **Page**

 b. Limiting Factors _____ 160

  (1) Reasonable Use _____ 161

  (2) The Accommodation Doctrine _____ 162

   (A) Rationale of the Accommodation Doctrine _____ 163

   (B) Elements of the Accommodation Doctrine _____ 163

  (3) For the Benefit of the Minerals Under the Surface _____ 165

   (A) Application of the Limitation _____ 165

   (B) Expanding the Implied Right _____ 167

  (4) In Accordance With Lease Terms and Applicable Statutes, Ordinances, Rules and Regulations __ 169

   (A) Restriction by Lease Provisions _____ 169

   (B) Restriction by Statutes, Ordinances, Rules, or Regulations _____ 170

Chapter 8. Essential Clauses of Modern Oil
            and Gas Leases—Continued   **Page**
  D. The Term Clause _____ 172
      1. The Primary Term _____ 172
      2. Secondary Term _____ 173
          a. The Meaning of "Production" 173
          b. How Much Production Is Re-
              quired: "Production in
              Paying Quantities" ___ 175
              (1) Determining   Operating
                  Revenues and Oper-
                  ating Costs _____ 177
                  (A) Operating   Revenues 177
                  (B) Operating Costs ____ 178
              (2) The Time Factor _____ 180
              (3) Equitable Considerations 181
  E. The Drilling-Delay Rental Clause ___ 181
      1. "Unless" v. "Or" Leases _____ 183
          a. The "Unless" Lease _____ 183
          b. The "Or" Lease _____ 185
          c. Forfeiture of "Or" Leases __ 186
      2. The Drilling Option _____ 189
          a. Commencement v. Comple-
              tion _____ 189
          b. In Good Faith _____ 191
          c. With Due Diligence _____ 191
      3. Payment of Delay Rentals _____ 193
          a. Protection of the Lessee by
              the Courts _____ 195
              (1) General Rule: Equity
                  Not Applicable _____ 195
              (2) Revivor of the Lease ___ 195
              (3) Estoppel and Waiver ___ 196

Chapter 8. Essential Clauses of Modern Oil and Gas Leases—Continued **Page**

    b. Protection of the Lessee by Lease Clauses ------- 198
       (1) Payment to an Agent -- 198
       (2) Notice of Assignment Provisions ----------- 200
       (3) Notice of Nonpayment Clauses ---------- 203
         (A) Enforceability of Notice of Nonpayment Clauses ---------- 203
         (B) Drafting Notice of Nonpayment Clauses ---------- 205
       (4) Use of "Or" Leases ----- 205
       (5) Use of Paid-Up Leases - 205
  F. Conclusion ---------------------- 208

Chapter 9. Defensive Clauses in Oil and Gas Leases --------------------- 209
  A. Clauses to Extend or Maintain the Lease --------------------- 210
    1. Dry Hole Clauses ------------ 210
      a. What Is a Dry Hole ------- 212
      b. When Is a Dry Hole Completed ----------------- 212
      c. When Payment Is Due ---- 212
    2. Operations Clauses ----------- 215
      a. Well Completion v. Continuous Operations Clauses 216

Chapter 9. Defensive Clauses in Oil and Gas
Leases—Continued **Page**

    b. Delay Between Completion of Operations and Production _____ 219

3. Pooling and Unitization Provisions _____ 220
    a. Community Leases _____ 222
    b. Pooling Clauses _____ 222
    c. Unitization Clauses _____ 224
    d. Problems Under Pooling and Unitization Clauses __ 227
        (1) Exercise in Good Faith _ 228
        (2) Time of Exercise _____ 229
        (3) Duty to Exercise the Power _____ 230
        (4) Cross-Conveyance Theory 231
        (5) The Right of a Non-Executive Owner to Ratify _____ 232
        (6) Conflict With the Rule Against Perpetuities _ 234
    e. Pugh Clauses or Freestone Riders _____ 234

4. Force Majeure Clauses _____ 236
    a. In General _____ 236
    b. Precise Terms Important __ 237

5. Shut-In Royalty Clauses _____ 238
    a. Scope of the Clause: Gas or Oil and Gas _____ 240
    b. Problems of Interpretation and Administration __ 241
        (1) What Is a "Shut-In" Well 241

Chapter 9.   Defensive Clauses in Oil and Gas
             Leases—Continued   **Page**
             (2) When Is a Well "Shut-In" 241
             (3) Shut-In for Reasons Oth-
                 er Than Lack of Mar-
                 ket _____ 242
             (4) How Long May Pay-
                 ments Be Made _____ 242
             (5) Effect of Failure to Pay 242
        6.   Cessation of Production Clauses 244
        a.   Temporary Cessation of Pro-
             duction Doctrine _____ 244
        b.   Lease Provisions _____ 245
   B.   Administrative Clauses _____ 248
        1.   Payment of Delay Rentals _____ 249
        2.   Warranty Clauses _____ 249
        3.   Lesser Interest Clause _____ 251
        4.   Subrogation Clause _____ 252
        5.   Equipment Removal Provisions 253
        6.   Notice of Assignment Clause ___ 254
        7.   No Increase of Burden Provisions 254
        8.   Separate Ownership Clause ____ 255
        9.   Surrender Clause _____ 255
       10.   Notice Before Forfeiture and Ju-
             dicial Ascertainment Clauses _ 256

Chapter 10.   The Lease Royalty Clause _____ 258
   A.   Common Royalty Provisions _____ 259
   B.   Nature of the Lessor's Royalty In-
        terest _____ 260
   C.   Deductions From Royalty _____ 261
   D.   The Market Value/Proceeds Problem 263
        1.   The Basis of the Dispute _____ 264

Chapter 10. The Lease Royalty Clause—Continued   **Page**
        2. Division of the Cases _____ 265
        3. How to Determine Market Value 267
        4. Curing or Avoiding the Market
            Value/Proceeds   Royalty
            Problem _____ 269
            a. Seeking   a   Compensating
                Higher Price _____ 269
            b. Specific Lease Language ___ 271
            c. Use of Division Orders ____ 272
   E. Failure to Pay Royalty _____ 274

Chapter 11. Implied Covenants in Oil and
             Gas Leases _____ 276
   A. The Basis of Implied Covenants _____ 276
        1. Implied in Fact or in Law? ____ 276
        2. Significance of the Distinction _ 277
   B. The   Reasonable   Prudent   Operator
            Standard _____ 279
        1. Competence _____ 280
        2. With Due Regard for the Lessor's Interests _____ 281
   C. Common Implied Covenants _____ 281
        1. The Implied Covenant to Test __ 282
        2. The Implied Covenant to Reasonably Develop _____ 283
            a. Elements of Proof of Breach 283
                (1) Probability of Profit ___ 284
                (2) Imprudent Operator ____ 284
            b. Stumbling Blocks to Enforcement _____ 286
                (1) Notice to the Lessee ___ 286
                (2) Disclaimer or Limitation
                    in the Lease _____ 287

Chapter 11. Implied Covenants in Oil and
Gas Leases—Continued **Page**
c. Remedies for Breach _____ 289
(1) Cancellation _____ 289
(2) Conditional Cancellation 290
(3) Damages _____ 290
3. The Implied Covenant for
Further Exploration _____ 291
a. Distinguished From the Covenant for Reasonable De-
velopment _____ 292
b. Elements of Proof _____ 295
c. Stumbling Blocks to Enforcement _____ 296
(1) Notice to the Lessee ___ 296
(2) Disclaimer or Limitation
in the Lease _____ 297
d. Remedies for Breach _____ 297
4. The Implied Covenant to Protect
Against Drainage _____ 298
a. Elements of Proof: In General _____ 300
(1) Substantial Drainage ___ 300
(2) Probability of Profit ___ 301
b. Elements of Proof: Where
There Is Drainage by the
Lessee _____ 301
c. Stumbling Blocks to Enforcement _____ 304
(1) Notice to the Lessee ___ 304
(2) Disclaimer or Limitation
in the Lease _____ 304
(3) Waiver or Estoppel ____ 306

Chapter 11. Implied Covenants in Oil and
             Gas Leases—Continued   **Page**
          d. Remedies for Breach _____ 306
              (1) Damages _____ 306
              (2) Cancellation or Condi-
                  tional Cancellation ___ 307
       5. The Implied Covenant to Market 307
          a. Within a Reasonable Time _ 308
          b. At a Reasonable Price _____ 309
          c. Proof of Imprudence _____ 309
          d. Remedies for Breach _____ 310
              (1) Failure to Market Within
                  a Reasonable Time ___ 310
              (2) Failure to Market at a
                  Reasonable Price _____ 311
          e. Stumbling Blocks to Enforce-
                  ment _____ 312
              (1) Notice to the Lessee ___ 312
              (2) Waiver or Estoppel ____ 312
              (3) Disclaimer or Limitation
                  in the Lease _____ 313
       6. The Implied Covenant to Operate
           With Reasonable Care and Due
           Diligence _____ 314
    D. The Future of Implied Covenants ___ 315

Chapter 12. Lease Transfers _____ 317
    A. Lessee's Right to Transfer _____ 317
    B. Effect of Transfer on the Lessor ____ 318
       1. Further Rights Against the
           Lessee _____ 318
       2. Future Rights Against the
           Transferee _____ 319
          a. Lease Covenants Run With
              the Land _____ 319
          b. The Assignment/Sublease
              Distinction _____ 320

Chapter 12. Lease Transfers—Continued   **Page**

C. Rights and Duties of the Lessee and His Transferee _____ 321

1. Protection of Non-operating Interests by Implied Covenants 321

2. Protection of Non-operating Interests Against "Wash Out" __ 322

3. Implied Covenants of Title From the Lessee _____ 323

4. Performance or Failure of Performance by a Partial Assignee _____ 324

5. Divisibility of Implied Covenants After a Partial Assignment __ 326

PART IV. TAX AND BUSINESS MATTERS _____ 328

Chapter 13. Oil and Gas Taxation _____ 328

A. Basic Principles _____ 328

1. The Property Concept _____ 329

2. Intangible Drilling Costs Deduction _____ 329

3. Depletion _____ 331

  a. Cost Depletion _____ 331

  b. Percentage Depletion _____ 332

    (1) Economic Interest Concept _____ 333

    (2) Independent Producers' and Royalty Owners' Limitation _____ 333

OUTLINE

Chapter 13. Oil and Gas Taxation—Continued
                                                         Page
B. Taxation of Transfers of Mineral Rights _____ 334
    1. Lease Payments _____ 334
       a. Lease Bonus _____ 334
       b. Delay Rentals _____ 335
       c. Royalty Payments _____ 335
    2. Lease Transfers _____ 335
       a. The Sale/Sublease Distinction _____ 335
       b. Production Payments _____ 337
       c. Sharing Arrangements _____ 338
C. Taxation of Oil and Gas Development 339
    1. Forms of Ownership _____ 339
       a. Corporation _____ 340
       b. Partnership _____ 341
       c. Concurrent Ownership _____ 341
    2. Taxation of Search Costs: Geological and Geophysical Costs 342
    3. Taxation of Development Costs 343
       a. Intangible Drilling Costs ___ 343
       b. Equipment Costs _____ 343
    4. Taxation of Oil and Gas Production _____ 344
       a. Tax Treatment of the Owners of Production _____ 344
       b. The Windfall Profit Tax ___ 344

Chapter 14. Oil and Gas Contracts _____ 346
A. Support Agreements _____ 347
    1. Purpose _____ 347
    2. Kinds of Support Agreements __ 347
B. Farmout Agreements _____ 348

Chapter 14. Oil and Gas Contracts—Continued
                                                              Page
 C. Operating Agreements ------------ 350
  1. Purpose --------------------- 350
  2. Common Substantive Issues --- 351
   a. Scope of the Operator's Authority ---------------- 351
   b. Initial Drilling ----------- 353
   c. Additional Development --- 353
 D. Drilling Contracts ---------------- 353
  1. Kinds of Drilling Contracts ---- 354
  2. Model Form Drilling Contracts - 355
 E. Gas Contracts -------------------- 356
  1. Common Substantive Issues --- 356
   a. Price -------------------- 357
    (1) Price Escalation ------- 357
    (2) Deregulation --------- 358
    (3) Buyer Protection ------- 358
  2. Take or Pay Provisions ------- 359
  3. Reserves Committed --------- 360
  4. Reservations ---------------- 360
  5. Conditions of Delivery -------- 361
 F. Gas Balancing Agreements -------- 361
  1. The Gas Balancing Problem --- 361
  2. Common Substantive Provisions 362
 G. Division Orders ------------------- 363

Glossary of Oil and Gas Terms ----------- 365

Appendix of Forms ---------------------- 389

Index ---------------------------------- 411

*

TABLE OF CASES

References are to Pages

Acker v. Guinn, 111, 112

Alexander v. Oates, 186, 188

Amoco Production Co. v. Alexander, 299, 314

Amoco Production Co. v. First Baptist Church of Pyote, 309, 311

Amoco Production Co. v. Underwood, 228

Atlantic Ref. Co. v. Beach, 119

Atlantic Ref. Co. v. Shell Oil Co., 194, 202

Baldwin v. Blue Stem Oil Co., 174, 190

Black v. Shell Oil Co., 125

Body v. McDonald, 128

Champlin Exploration, Inc. v. Western Bridge & Steel Co., 11

Clifton v. Koontz, 176

Cook v. El Paso Natural Gas Co., 303

Corlett v. Cox, 120

Cotiga Development Co. v. United Fuel Gas Co., 291

Duhig v. Peavy-Moore Lumber Co., 127, 128, 129, 130, 131, 132, 251

Elliff v. Texon Drilling Co., 13

Exxon Corp. v. Middleton, 265, 268, 273

Federal Land Bank of Houston v. United States, 96

FERC v. Pennzoil Producing Co., 270

First Nat'l Bank in Weatherford v. Exxon Corp., 269

Gard v. Kaiser, 244

Gerhard v. Stephens, 25

TABLE OF CASES

Getty Oil Co. v. Jones, 162, 163, 164
Gilbertson v. Charlson, 129, 130
Girolami v. Peoples Natural Gas Co., 186
Globe v. Goff, 191
Greer v. Stanolind Oil & Gas Co., 194
Gulf Production Co. v. Kishi, 288
Gulf Ref. Co. v. Shatford, 200

Hartman v. Potter, 129, 130
Henry v. Ballard & Cordell Corp., 266
Hillard v. Stephens, 266
Hoffman v. Magnolia Petroleum Co., 136
Holloway's Unknown Heirs v. Whatley, 159
Hoyt v. Continental Oil Co., 247
Humble Oil & Ref. Co. v. Harrison, 196, 197, 200
Humble Oil & Ref. Co. v. Kishi, 45

Lewis v. Grininger, 203, 205
Lightcap v. Mobil Oil Corp., 266, 268

McEvoy v. First Nat'l Bank & Trust of Enid, 103
McMahon v. Christmann, 131, 251
Mitchell v. Amerada Hess Corp., 293
Montana Power Co. v. Kravik, 266

Northern Natural Gas Co. v. Grounds, 155

Paddock v. Vasquez, 136, 137
Phillips Petroleum Co. v. Curtis, 193
Piney Woods Country Life Sch. v. Shell Oil Co., 266
Prairie Oil and Gas Co. v. Allen, 81, 83, 84, 85

Reed v. Wylie, 112, 113, 114
Ruiz v. Martin, 232

Saulsberry v. Siegel, 245
Sinclair Oil & Gas Co. v. Masterson, 289
Smith v. Allison, 159
Spell v. Hanes, 125
Stewart v. Amerada Hess Corp., 179

TABLE OF CASES

Sun Oil Co. v. Whitaker, 163, 164
Sunac Petroleum Corp. v. Parkes, 217
Superior Oil Co. v. Jackson, 202
Superior Oil Co. v. Stanolind Oil & Gas Co., 214, 215
Sword v. Rains, 219, 220, 238

Tara Petroleum Corp. v. Hughey, 266
Texaco, Inc. v. Short, 89
Texaco, Inc. v. Fox, 180
Texas Oil & Gas Corp. v. Vela, 267

Warm Springs Development Co. v. McAulay, 182
Waseco Chemical & Supply Co. v. Bayou States Oil Corp.,
    285, 290, 314
West v. Alpar Resources, Inc., 266
Wettengel v. Gormley, 141
Wooley v. Standard Oil Co., 204

Young v. Jones, 194

*

OIL AND GAS LAW
IN A NUTSHELL

*

PART I

THE NATURE AND PROTECTION OF OIL AND GAS RIGHTS

CHAPTER 1

THE FORMATION AND PRODUCTION OF OIL AND GAS

Oil and gas are the liquid and gaseous forms of petroleum, a chemically complex substance composed of hydrogen and carbon with trace amounts of oxygen, nitrogen, and sulphur. Petroleum occurs in gaseous, liquid, and solid states, depending upon its physical composition, temperature and pressure.

A. FORMATION OF OIL AND GAS

Petroleum is found in sedimentary rocks associated with ancient seas. The generally accepted theory for its origin is that sediment from rivers and remains of marine plants and animals simultaneously accumulated on sea floors, forming layer upon layer of sediment and organic residue. As layers were buried deeper and deeper, they were compressed and subjected to increasing

[1]

pressure from the overlaying sediment. Increased pressure generated heat which, acting
upon the sediment for tens to hundreds of millions of years, transformed the organic material
into crude oil and natural gas.

Originally, sediment deposits laid down on the
sea floor were nearly horizontal. However, millions of years of deformation of the earth's crust
left the layers folded and faulted, forming trap
geometries where petroleum might accumulate.
These "anomalies" are of limited size and can occur from depths of several hundred feet to tens
of thousands of feet. The oil and gas industry
tries to locate those anomalies by mapping, rock
evaluation, and seismic studies.

All sedimentary rocks contain pore spaces (porosity) between the sediment particles. However,
in some rocks the pore spaces are interconnected
(permeable), and in other rocks the pore spaces
are isolated from one another (impermeable). Because petroleum is lighter than water, it will float
upwards through the original sea water contained
in the pores of permeable rocks. It is common to
find layers of permeable rocks bounded by layers
of impermeable rocks. Whenever the geometry
of the rock layers is described by a "dome-like"
shape, a petroleum trap is formed. A petroleum
"reservoir" is defined when the pores of the rock
in the "trap" contain commercially produceable
quantities of oil or gas. As the diagram shows, it
is reservoirs that the oil industry seeks to locate
by drilling operations.

[*2*]

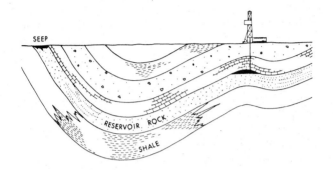

B. DRILLING FOR OIL AND GAS

To test an anomaly, a well has to be drilled. There is no known technology to determine the presence of petroleum in rocks thousands of feet below the earth's surface without drilling. Two types of drilling rigs are used—the rotary rig and the cable tool rig. The cable tool rig is an older style, rarely used today, that is based upon the principle of raising and letting fall a heavy bit to pulverize the rock. This technique, though inexpensive, is limited in application and cannot be used below depths of a few thousand feet.

Most wells, both onshore and offshore, are drilled by the rotary technique. A typical rotary drilling rig consists of a derrick structure, a string of pipe, a drill bit, circulating fluid, and a derrick floor rotary turntable, as the following drawing shows.

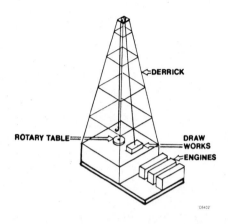

The derrick structure is assembled over the drill
site. The drill bit is screwed into the bottom of a
thirty-foot length of pipe which passes through
the rotary table. The circulating fluid (a chemically complex mud or air) is forced down the inside of the pipe, out through the jets in the bit,
and back to the surface, where the cycle is re-
peated. As the turntable rotates, the drill pipe is
forced to rotate, which forces the drill bit to rotate against the rock and abrade the rock into
matchhead-size pieces, which the circulating mud
carries to the surface. A drawing of a rotary
drill bit follows.

[C6403]

Every time a thirty-foot depth of hole is drilled, a new stand of pipe is added. At a depth of thirty thousand feet the continuous drill string would consist of 1000 pieces of pipe screwed together end-to-end. When the total depth recommended is reached, the well data is evaluated and a decision made to (1) plug and abandon, if there are no indicators of petroleum in the rock, or (2) "set pipe" if petroleum is present.

C. PRODUCING OIL AND GAS

If it appears that a well contains commercial quantities of oil or gas, a continuous string of production casing pipe (thirty-foot lengths screwed together) is placed in the hole. Cement is forced around the outside of the pipe, sealing off the space between the rock wall and the pipe exterior. A perforating sonde is lowered down the hole inside the pipe to the depth of the potential petroleum-bearing rock. The sonde contains explosives that penetrate the casing, cement, and several inches of rock. This allows the petroleum in the rock to drain into the well bore. Sometimes it is necessary to "stimulate" the well by forcing fluids into the rock to fracture it and to inject acid to dissolve away some of the rock. Both procedures will improve permeability.

If the natural pressure within the rocks is high, oil will "flow" to the surface of its own accord. If the pressure is low, pumping equipment will be installed to "lift" the oil to the surface. Gas will normally flow, and the system of valves and gauges at the top of the well which control its flow is called a "Christmas tree." Crude oil occasionally will flow, but generally the familiar "horsehead" pumping jack is required.

Because crude oil, natural gas, and salt water are commonly produced simultaneously, it is necessary to flow fluids produced through a separator to remove the natural gas and perhaps through a heater-treater to separate oil and water. The oil and gas are then stored, piped, or trucked to the refinery. The diagram that follows shows a typical production configuration.

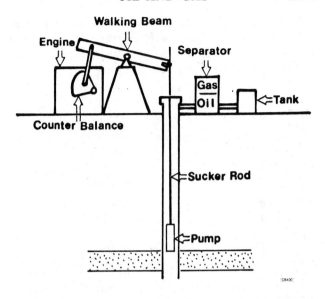

After some period of time, ranging from several months to many years, the natural or "primary" pressure in the reservoir rocks may deplete to such a level that petroleum will no longer flow into the well bore. At that time the operating company will evaluate the economics of artificially enhancing the reservoir pressure by injecting water or gas into the reservoir. In more advanced enhanced recovery projects, fire flooding and complex chemical techniques are used to improve reservoir pressure. More than 30% of U.S. production today comes from enhanced recovery techniques.

CHAPTER 2

OWNERSHIP OF OIL AND GAS RIGHTS

A. THE AD COELUM DOCTRINE

Oil and gas law is an example of development of a new body of law by modification of existing common law concepts. In 1859, when the first oil well was drilled at Titusville, Pennsylvania to a depth of 69 feet, the prevailing principle of ownership was *cujus est solum, ejus est usque ad coelum et ad inferos,* commonly referred to as the *"ad coelum"* doctrine. Translated, that meant that the owner of the bundle of rights that we call "ownership" of property owned everything from the heavens above the surface of his land to the core of the earth beneath it.

It quickly became apparent that this common law principle, which works well with "hard" minerals, was not appropriate to govern extraction of petroleum. Oil and gas are fugacious; they may move from place to place within sedimentary rock. In addition, oil and gas are fungible; it is difficult to determine whether a given MCF of gas or a barrel of oil produced has been drawn from under one tract of land or another.

Adherence to the *ad coelum* principle would have hamstrung the development of an industry potentially important to America's continued economic development. It would have discouraged mineral owners from drilling for fear of incurring liability for drainage from their neighbors' property. Further, application of the *ad coelum* principle would have conflicted with *laissez faire*, the prevalent political and economic theory of the era, that emphasized a policy of rewarding the diligent to the ultimate benefit of society. It is not surprising that the *ad coelum* doctrine was modified by the rule of capture.

B. THE RULE OF CAPTURE

The rule of capture has been described as a "rule of convenience." What is meant is that the rule developed from recognition by the courts of society's need for energy resources rather than from the logic of earlier precedents. It was a substantial departure from the principle of law governing the extraction of hard minerals to permit development of oil and gas resources. The rule of capture may be stated as follows:

The owner of a tract of land acquires title to the oil and gas that he produces from wells drilled thereon, though it may be proved that part of such oil and gas migrated from adjoining lands. There is no liability for capturing oil and gas that drains from another's lands.

The rule of capture is unusual as a rule of law because it is a rule of nonliability. It gives the owner of minerals the shield of a positive legal principle as he develops the oil and gas resources of his land. So long as he conducts his operations without trespassing or interfering with the rights of neighboring owners to drill to the same formation under their lands he will not be liable. All the oil or gas he captures will belong to him, even if it is drained from beneath others' lands.

C. LIMITS TO THE RULE OF CAPTURE

The rule of capture is not a perfect shield to liability, however. There are inherent limitations, limitations imposed by the doctrine of correlative rights, and statutory limitations. All are consistent with the underlying rationale of the rule of capture.

1. INHERENT LIMITATIONS

The rule of capture is inherently limited by its rationale. As has been noted, it developed from recognition by the courts that mechanical application of the *ad coelum* doctrine would deprive society of important energy resources. The rule was developed to encourage development of oil and gas resources for the benefit of society. The purpose limits its protection.

a.   Escaped Hydrocarbons

The inherent limitation of the rule of capture has been demonstrated by several cases that have raised the issue of the right of a mineral interest owner to capture oil or gas previously produced by another. A distinction has been made between natural gas as it exists in its natural state and extraneous gas which is captured elsewhere and injected into storage. Gas in its natural state is subject to capture. But once captured, it remains the property of the one who captured it until abandoned. For example, in *Champlin Exploration, Inc. v. Western Bridge & Steel Co. Inc.*, 597 P.2d 1215 (Okl.1979), the Oklahoma Supreme Court held that the rule of capture did not protect a mineral interest owner that dug trenches on its premises and pumped out refined hydrocarbons that escaped from a nearby refinery and gathered in the trenches. Except in Kentucky, similar results have been reached by the relatively few courts that have considered situations in which natural gas produced and injected into a storage reservoir is recaptured in a well drilled on a tract not subject to the storage rights. Generally, that result has been reached by reasoning that oil or gas becomes personal property when produced, so that ownership is not lost by mere loss of possession. The result is consistent with the purpose underlying the rule of capture because the activity for which protection of the rule was sought would have added nothing to society's energy supplies.

b. Drainage by Enhanced Recovery Operations

Limitations upon the rule of capture have also been recognized in situations in which capture has been brought about by enhanced recovery operations, procedures that improve the productive

[*11*]

capacity of the reservoir by injecting fluids to increase the pressure differential or to move oil and
gas in place to the borehole. Here, the courts
have generally recognized that permitting a lessee to sweep oil and gas from under property of a
neighbor by use of water flooding techniques is
beyond the scope of the rule of capture. A few
courts, including courts in Oklahoma and Arkansas, have held that the mineral owner conducting
secondary recovery operations is liable to adjoining mineral owners who are drained of oil and
gas on a theory of nuisance or trespass. Others,
including courts in Texas, Nebraska, and Mississippi, have rejected liability for nuisance or trespass where the drained party has refused what
the court has considered to be a "fair" proposal
to participate in an enhanced recovery program
and the state conservation agency has approved
the project as necessary to prevent waste and
maximize production. Though such decisions
find no liability, they recognize that the rule of
capture will not excuse all drainage.

The rule of capture should be applied to protect
mineral interest owners engaged in enhanced recovery operations, at least where those operations are necessary to maximize ultimate production. In any oil and gas reservoir, there are
reserves that can be produced by primary recovery techniques and those which can be produced
only through the use of secondary or tertiary recovery techniques. Use of enhanced recovery
techniques to sweep away reserves recoverable
by an adjoining property owner by primary recov-

ery techniques should not be protected by the rule of capture. Those reserves would have been produced anyway; application of enhanced recovery techniques merely speeds up their production and permits one mineral interest owner rather than another to produce them. On the other hand, when it is found by the conservation agency that enhanced recovery techniques permit production of oil and gas that would not be produced by primary production techniques, the interest of society in maximizing production of its resources dictates that the activity be protected by the rule of capture. However, a requirement that the drained owner be offered an opportunity to participate in the unit operation on a fair basis should be imposed to prevent enhanced recovery from being used as a weapon against other owners' correlative rights.

2. DOCTRINE OF CORRELATIVE RIGHTS

Another limitation to the rule of capture is the doctrine of correlative rights, illustrated in the classic case of *Elliff v. Texon Drilling Co.*, 146 Tex. 575, 210 S.W.2d 558 (1948). There, Texon's negligence permitted one of its wells on property adjoining that of Elliff to blow out and burn, causing large quantities of oil and gas to be drained from under Elliff's property. When Elliff sued for damages for the lost oil and gas, Texon raised the rule of capture as a defense. The Texas Supreme Court rejected the defense, noting that each owner has a right to a fair and

[*13*]

equitable share of the oil and gas under his land as well as the right to be protected against negligent damage to the producing formation. This is the correlative rights doctrine. Because Texon was wasting the oil and gas rather than selling or using it, the rule of capture did not shield it from liability.

The correlative rights doctrine is a corollary to the rule of capture and it follows from its logic. Since the rule of capture was adopted to benefit the public interest in plentiful energy by making it possible to develop oil and gas resources, activity not consistent with that purpose is not protected by the rule of capture. Waste or wasteful production techniques will bring liability. So will negligent damage to the ability of the producing formation to produce for others. Positively stated, the correlative rights doctrine provides that each owner of minerals in a common source of supply has the right to a fair chance to produce oil and gas from the reservoir substantially in the proportion that the quantity of recoverable oil and gas under his land bears to the quantity in the reservoir.

3. CONSERVATION LAWS

Neither the doctrine of correlative rights nor the inherent limitations discussed above have provided sufficient limits to the rule of capture. The problem can be illustrated if you imagine that you own a tract of 640 acres that you are advised can be efficiently drained by a single well located

[*14*]

anywhere on the tract. If the rule of capture and the doctrine of correlative rights are the legal rules applicable, where will you drill your well? And, will you drill a single well or several?

The answer to the first question is that, if you are astute, you will drill your first well as close to the boundary of your tract as you can, rather than in the center. Your motivation will be to use the rule of capture to drain oil and gas from your neighbor's land as well as from your own. Even if you are not motivated by greed for the oil and gas under your neighbor's land (as you ought to be if you are a reasonable economic person), you will drill your first well close to the boundary line to protect yourself against drainage from a well drilled on your neighbor's property.

Furthermore, whether your motive is greed or a desire for protection, you will probably drill not one but several wells along your boundary. If you do not, you will not gain the maximum advantage from the rule of capture and you will leave yourself exposed to a neighbor who drills on his property close to your boundary.

a.   Economic and Physical Waste

What is wrong with the scenario outlined? The problem is that it leads to physical and economic waste. The economic waste is easy to see. Because your neighbors will have the same legal rights and economic motivation as you, over a period of time there will tend to be many more wells drilled in the area than are necessary to drain it efficiently. Each owner will drill the number of wells he judges necessary to maximize his benefits or to protect himself. And each of the owners will be pressed by economics to drill as many wells as the most active neighbor. If one does not, he will be at a disadvantage. The process will be economically wasteful because it will be more costly than necessary due to the drilling of more wells than are required to drain the field efficiently.

Economic waste from over-drilling is likely to lead to physical waste. Once wells are drilled, each owner will feel compelled to produce them as fast as possible to drain the reservoir before his neighbors and to increase his chances of recovering his costs of drilling and maximizing his profits. But short term over-production from the reservoir is likely to result in long term total recovery of a percentage of the oil and gas in place *less* than what might be achieved by slower production. The natural expansion of oil and gas toward the borehole will be dissipated among many boreholes with the result that much of the oil and gas in place will be left in the formation. In addition, over-production of oil and gas is likely to push down the price for which oil and gas produced can be sold. In the early years of the century, that happened often. As the price dropped, wasteful uses proliferated; e.g., the use of natural gas for lamp black. When the price dropped below the point that operating revenues exceeded operating costs, wells were plugged and abandoned and the remaining petroleum was "locked in" the formation.

The correlative rights doctrine does not prevent the physical and economic waste inherent in application of the rule of capture because of the problem of proving causation. Over-drilling by one owner is not the direct cause of waste of oil or gas or of damage to the producing formation. Over-drilling by all of the owners as a group is the direct cause of waste. The doctrine of correlative rights gives each individual rights against other individuals, and it cannot be easily applied when all are at fault.

b. Function of Oil and Gas Conservation Laws

As the problems in unrestrained application of the rule of capture became apparent, states began developing petroleum conservation laws. Today, such laws are the keystone of the legal structure governing oil and gas development.

(1) Purpose

The purpose of oil and gas conservation statutes is to avoid physical and economic waste of oil and gas resources. However, they are concerned not only with saving resources but also with encouraging their rational development. Rational development is a part of the prevention of waste because it maximizes ultimate recovery. Conservation laws seek to further the public's interest in conservation and rational development and to protect the correlative rights of operators and landowners.

(2) Well Spacing Rules

A typical conservation statute consists of a series of balancing provisions that seek to maintain an equilibrium between public interest and private rights. The most important of those provisions deals with well spacing. Since the primary problem with the rule of capture is that greed and the need for protection leads owners of oil and gas rights to drill wells too close together, an important step toward control of the problem is to set rules for the spacing of wells, requiring that wells be spaced specified distances from boundary lines and from one another so that excessive overlapping drainage will not occur. Spacing rules prevent over-drilling by limiting the number of wells that can be drilled in a given area.

(3) Production Regulation

Analogous provisions extend regulation to production. *Gas-oil ratio rules* and *water-oil ratio rules* permit control of the production ratios between oil and gas and oil and water in order to prevent dissipation of the pressure that produces oil and gas. For example, if pressure is lost from a gas drive reservoir by flaring natural gas at a time when there is substantial demand for oil but little demand for natural gas, the oil may remain trapped in the formation forever. Gas-oil ratios and water-oil ratios require that operators cease production when ratios of gas to oil or water to oil determined appropriate are exceeded. Thus, they prevent waste of petroleum and enhance actions by producers to maintain pressure.

Another kind of production regulation is establishment of *production allowables*. Production allowables, sometimes called *prorationing rules*, put daily, weekly or monthly limits on production of oil and gas to prevent overproduction. Where the limits are imposed to prevent overly fast production of a reservoir (in excess of the scientifically determined maximum efficient rate) the process is sometimes called *MER prorationing*. Where the production allowables are used to prevent wide fluctuations in the price of oil, as was the practice of some of the southwestern producing states prior to the world oil shortage of the 1970's, the process may be referred to as *market demand prorationing*.

Limits on production may also be set by provisions for compulsory unitization or encouragement of voluntary unitization. *Unitization* refers to the joint operation of some or all of the wells over a producing formation to maximize the production from the unit rather than from any individual well. Unit operations may substantially increase the percentage of oil and gas recovered in the long run. In most instances, unitization could not be successful without the help of compulsory unitization laws because many landowners and producers see a short term advantage to individual production.

(4) Protection of Correlative Rights

Inevitably, conservation provisions aimed at maximizing ultimate production and preventing waste interfere with the correlative rights of some owners of mineral rights. Suppose, for example, that the well spacing statute requires square 640 acre spacing units and provides further that the location of the well be approximately in the middle of the unit. Unless statutory relief of some kind is given, the owner of the minerals under the northeast 40 acre square of the tract has lost his correlative rights because state law will bar him from drilling. To protect the correlative rights of the small tract owner adversely affected by spacing rules, conservation statutes typically include compulsory pooling provisions or exception tract provisions.

Compulsory pooling provisions provide the mineral owner whose correlative rights are threatened by potential drainage with a legal right to participate in the unit. They may also be used by mineral owners who want to develop their properties to force recalcitrant owners to participate or to lease on fair terms. Generally, such statutes give any owner of mineral interests in property the right to protect himself against drainage by demanding compulsory pooling. Compulsorily pooled owners may be given the choice of (1) agreeing to participate in drilling and pay their share of the costs, (2) agreeing to give up their operating rights in return for a bonus payment and royalty determined appropriate by the state conservation agency, or (3) electing to be "carried" for drilling and completion costs. Electing to be carried means that the other owners will advance costs of drilling and completion, but the carried party will receive only a royalty on production until the parties advancing the costs have recovered some multiple of their investment, after which the carried party will "back in" to a share of the working interest in addition to or in place of the royalty. Whatever the precise terms of the compulsory pooling statute, it makes it possible for all who own mineral rights within the unit to share in production.

Exception tract statutes protect correlative rights of small tract owners by permitting them to drill wells on their properties even though their land is not large enough or the proper shape to meet the provisions of the spacing unit statute. They permit exceptions to be made to the spacing rules to guarantee small tract owners the right to drill. In Texas, until the early 1960's the Railroad Commission (the Texas conservation agency) granted owners who drilled wells on exception tracts a production allowable sufficient to permit them to recover their costs plus a reasonable profit. Eventually, it was recognized that this practice was little more than a license to drain oil and gas from under other properties that seriously conflicted with well spacing rules. The practice was discontinued, and production allowables on exception tract wells were set to permit the owners to recover the oil and gas under their property, without reference to whether production would be economically profitable. Such a limitation is universal among states that have exception tract statutes today.

Oil and gas conservation statutes are an attempt to protect both the public and private interests in oil and gas development. They were the first of many applications of the state's police power to the oil and gas industry. They are an effective and important limit upon the historical problems of the rule of capture.

D. THEORIES OF OWNERSHIP OF OIL AND GAS

1. NON–OWNERSHIP AND OWNERSHIP IN PLACE THEORIES

While the judges that developed the rule of capture were willing to apply a rule other than the *ad coelum* doctrine, they sought to justify their departure from precedent by distinguishing ownership of oil and gas from other substances found in the earth. The earliest courts to face the problem analogized to the law of wild animals. They held that the owner of oil and gas rights did not own oil and gas until they had been subjected to his control by capture in his well. Prior to that time, what the owner of oil and gas rights had was an exclusive right to explore for, develop and produce oil and gas from the premises subject to his rights. Many courts have characterized such rights as a *profit a prendre*, a right to go on the land and take some part of the land or a product of it.

Other courts took an approach closer to the *ad coelum* principle. They rationalized that since oil and gas were a part of the soil, they were owned in place by the owner of the land in addition to the exclusive right to explore for, develop, and produce. The ownership was limited, however, by the fugacious nature of oil and gas; the ownership right to a particular barrel of oil or MCF of gas would be terminated if the oil and gas migrated to the land of another. These courts characterized the rights of an owner of oil and gas as a *fee simple determinable*, an ownership estate that terminates automatically upon stated occurrences.

The first rule is often referred to as the "nonownership" theory. It is followed in Oklahoma, Louisiana, California, and Wyoming as well as several other less prolific oil producing states. The second approach is often referred to as the "ownership in place" theory. Texas, New Mexico, Colorado, and Kansas are among the states that have adopted this theory of ownership. The courts in some states have never addressed the issue while those in others have addressed it inconsistently.

The rule of capture underlies both theories of ownership. It is inherent in the non-ownership theory; the only way ownership of oil and gas can be obtained in a non-ownership theory state is by capturing them. It is a caveat to the ownership in place theory; in an ownership in place theory state, the owner of oil and gas rights owns the right to oil and gas in place, subject to the right of others to divest him of his ownership by capturing it.

2. SIGNIFICANCE OF THE THEORIES OF OWNERSHIP

a.   The Corporeal/Incorporeal Distinction

The primary significance of the ownership theory embraced has followed from the recognition or non-recognition of the owner's present right of possession of oil and gas under his property. At common law, rights to land are classified as corporeal or incorporeal, according to whether they carry with them the right of physical possession. If an interest in land includes the right of possession of the land, it is classified as a corporeal ("of substance") right. On the other hand, if it includes only the right to use the land, it is classified as incorporeal ("not of substance"). By this analysis, rights to oil and gas are incorporeal in states embracing the non-ownership theory and corporeal in those that have adopted the ownership in place theory.

(1) Abandonment of Oil and Gas Interests

One important difference between corporeal and incorporeal rights at common law is that in-

[*24*]

corporeal rights can be abandoned, while corporeal rights cannot. The distinction has been applied in disputes over oil and gas rights. In the classic case of *Gerhard v. Stephens*, 68 Cal.2d 864, 69 Cal.Rptr. 612, 442 P.2d 692 (1968), the successors in interest to two corporations dissolved in 1915 sued to quiet title to oil and gas rights leased by Stephens to Shell in 1956 and thereafter profitably developed. The California Supreme Court held that oil and gas rights were subject to abandonment in California, noting that an earlier case had rejected ownership in place theory as the law of California. The court reasoned that oil and gas rights were a *profit a prendre*, an incorporeal right subject to loss by abandonment. By the analysis of *Gerhard v. Stephens*, oil and gas mineral rights, leasehold rights and royalties are all subject to abandonment in a state following the non-ownership theory. By definition, no owner of any oil and gas right can have the right to present possession of the oil and gas in place in a non-ownership theory state.

In states following the ownership in place theory, some oil and gas interests may be subject to abandonment. An oil and gas mineral right is an estate in the oil and gas in place. As such, it bestows upon its holder a present right of possession and it may not be abandoned. A leasehold interest, on the other hand, may be either a grant of the lessor's right to use the land to search, develop, and produce *and* his present right of possession in place, or it may be a grant only of the right to search, develop, and produce. The courts in some states that subscribe to the ownership in place theory take the position that whether the interest created in a lease is corporeal or incorporeal depends upon the language used in the lease granting clause; that is, whether or not the grant includes the minerals in place. Others, including Texas, have held that a lease ordinarily severs all of the grantor's mineral rights, including the present right to possession, whatever the language used in the grant. A royalty interest, a right to a share of oil and gas produced free of costs of production, should always be classified as an incorporeal right, regardless of the theory of ownership of the jurisdiction. A royalty is a right to oil and gas if and when it is produced. Therefore, there is no present right of possession.

Even in non-ownership theory states, the fact of abandonment is rarely found. Except in California, this writer knows of no reported cases finding that oil and gas rights have been abandoned. The reason is that abandonment requires a showing both of an extended period of non-use and an intention to abandon. Because oil and gas rights are the kind of property interest that one ordinarily may hold for an extended period without development and because development is the only way to use them, it is difficult to find evidence of intention to abandon. The courts have uniformly held that non-use by itself does not establish the requisite intent, though the length of the non-use may be considered in determining the intent. The California courts alone have been willing to consider the length of the non-use in conjunction with the existence of economic conditions that would make future use of the right unlikely in order to find the requisite intent. Therefore, except in California, application of the common law doctrine of abandonment to oil and gas rights in a non-ownership theory state is of little practical significance.

(2) Forms of Action to Protect Oil and Gas Rights

Classification of oil and gas rights as corporeal or incorporeal may affect the forms of action available to protect rights. Certain forms of action, such as trespass, ejectment and compulsory partition, were said by the common law courts to be possessory in nature; that is, they were available only to those who owned possessory interests in property. The courts of some states have recognized that distinction and have said that only holders of corporeal rights are entitled to possessory remedies. In those states, if the ownership in place theory has been adopted, then the owner of oil and gas rights is entitled to the possessory remedies. On the other hand, if the non-ownership theory has been adopted, he is not.

In as many jurisdictions as not, however, the distinction between corporeal and incorporeal rights as a prerequisite to maintenance of a possessory action has been ignored either by treating oil and gas rights as *sui generis* or by finding that codification of common law remedies has modified their application. Even in those states

[27]

that have maintained the common law distinction, the substantive rights of oil and gas owners have been protected by other remedies. For example, where ejectment is unavailable, an action to quiet title may generally be maintained.

b.   Classification of Oil and Gas Rights as Real Property or Personal Property

Other important distinctions involving the nature of rights to oil and gas have *not* followed from the ownership theory adopted by the jurisdiction. Classifying oil and gas rights as real property or personal property is one such distinction, though occasional cases have suggested otherwise.

For a variety of reasons, it may be crucial to the parties whether an interest in oil and gas is classified as realty or personalty. Intestate or testamentary rights may turn on the distinction. So may the applicable schedule of taxes, though taxing statutes usually are specific in coverage.

There is no correlation between the ownership theory embraced by the jurisdiction and classification of oil and gas interests as real property or personal property. The distinction between the ownership theories is whether or not the owner has a present possessory right to the oil and gas in place. The distinction between real property and personal property at common law turns on the duration of the interest rather than the possessory quality of it. If the interest's duration is that of a freehold estate—a life estate or a fee estate—it is real property; otherwise it is personalty. By the logic of the common law, any oil and gas interest, whether a mineral right, a leasehold right, or a royalty, is an interest in land. Whether that interest is classified as real estate or personal property ought to depend upon its duration. Interests that are for "life or longer" should be classified as real property. Interests with a lesser duration should be classified as personal property. Therefore, most oil and gas rights created (e.g. "for ten years and so long thereafter as oil and gas are produced . . .") are logically real property because they are of potentially perpetual duration. Of course, oil or gas, themselves, become personal property when produced by an exercise of the rights.

Whether a particular jurisdiction has classified oil and gas rights as real property or personal property is generally determined by statutory interpretation rather than by application of common law principles. Thus, in many states, perpetual mineral interests are not "real property" for purposes of the taxing statutes or the judgment lien statutes although they would be so classified by application of the common law standard.

c. Practical Impact of the Theory of Ownership

The theory of ownership embraced by a particular state is likely to be of more importance to law professors than to mineral interest owners.

The difference is in the nature of the interests that follows from the ownership. The theories may be of importance in specific fact situations. But on a day-to-day basis, the similarities are far greater and far more important than the differences.

CHAPTER 3

KINDS OF OIL AND GAS INTERESTS

Though oil and gas interests are generally created and conveyed like real property interests, the names given to the interests created are likely to be unfamiliar even to those familiar with real property transactions. Also, some of the characteristics associated with certain interests are peculiar to oil and gas law. In this chapter we will consider commonly encountered oil and gas interests and their characteristics.

A. FEE INTEREST

People working in the oil and gas industry frequently talk about the "fee interest" in property. By this they mean ownership of both the surface and the mineral rights in fee simple absolute.

Technically, references to the "fee interest" are incorrect. The word "fee" is used in property law to describe an estate or interest of inheritance, one that may be passed from generation to generation. It refers to the potential *duration* of an estate or interest rather than to the ownership rights it encompasses. An interest in the surface or minerals alone (at least in ownership in place theory states) may be held "in fee." However, the use of the term is well-established in the oil and gas industry to mean the whole "bundle of sticks" of rights in real property.

B.   MINERAL INTEREST

It is axiomatic in Anglo-American law that the owner of property rights can transfer property rights in whole or in part. Where the transfer is of less than the whole bundle of property rights owned, it is said that there has been a *"severance"*.

It has become common in the United States for mineral rights to be severed from the surface rights in land. Sometimes, severance is by a reservation or exception in a deed transferring the remaining interest. On other occasions, severance is by a direct grant of the mineral interest. A typical mineral deed is included in the Appendix. A severance may divide ownership of mineral rights and surface rights. Or, it may divide ownership of various kinds of minerals.

1.   MINERAL INTEREST INCLUDES RIGHT TO USE THE SURFACE

A mineral interest is more than just ownership of minerals. Whether the mineral interest is severed by deed or by reservation, it includes an implied easement to use the surface in such ways and to such an extent as is reasonably necessary to obtain the oil and gas under the property. Of necessity, mineral ownership implies a right to use the land surface over the minerals because mineral ownership would be economically value-less without access. An implied easement burdening the surface and benefitting the minerals is recognized on the basis either that it was the intention of the parties or that there is a public policy in favor of making property economically useful.

The right to use the land surface is so central to the ownership of mineral interests in oil and gas that most definitions of mineral rights are phrased in terms of surface use. A common definition follows:

> The mineral interest in oil and gas is the right to search for, develop and produce oil and gas from the described premises and, in states that have adopted the ownership in place theory, the present right to possess the oil and gas in place under the property.

As a practical matter, the right recognized in some states to present possession of the oil and gas in place is of little consequence. The only certain method of determining whether oil or gas are present is to drill a well to test the property. The easement for surface use holds the real economic value of the mineral interest.

2. CHARACTERISTICS OF MINERAL INTEREST OWNERSHIP

The economic reality, as well as the essential similarity from state to state, can be seen from the incidents of mineral interest ownership as they pertain to oil and gas. There is general agreement that mineral interest incidents fall into three classes:

a. *The easement for surface use*—the mineral interest owner has the right to use the surface of the land under which he owns the minerals to search for, develop and produce the minerals. Of course, the mineral interest owner must pay the costs of his efforts, and his easement to use the surface is limited by a standard of reasonableness and an obligation to accommodate the uses of the surface owner, if possible.

b. *The right to lease or sell the mineral interest*—the mineral interest owner has the right to explore for or develop oil and gas on the property in which he owns the mineral interest. He also can transfer that right to another. The right to lease is often referred to as the *executive right,* particularly when it is severed from the rest of the mineral interest.

c. *The right to benefits under an oil and gas lease*—because the mineral interest owner has the right to develop, he also has the right to whatever benefits are provided to the lessor under the terms of a lease that transfers the right to develop. Typically, these include the right to any payments made to induce the signing of the lease (bonus), any payments for maintaining the lease without development (delay rentals) or production (shut-in royalty) and any share of production allocated to the lessor (royalty).

Just as the owner of the fee interest may convey or reserve the mineral interest separately from the remainder of the property, so the owner of the mineral interest can separately convey or reserve some of the incidents of mineral ownership. O, the owner of the mineral interest in Blackacre, might convey to A the mineral interest reserving to himself the right to lease the property, the right to one half of any bonus, and the right to one quarter of any royalties. In such an event, both O and A would have something less than a "true" mineral interest. The flexibility afforded property owners in carving out unusual groupings of rights gives rise to frequent interpretative problems, some of which will be considered in Chapter 7.

3. LOUISIANA'S MINERAL SERVITUDE

Under Louisiana's civil law regime, the owner of land does not own fugacious minerals, so oil and gas rights cannot be severed from surface rights. However, a *mineral servitude* may be imposed upon land giving its holder the right to search for, develop and produce oil and gas. A mineral servitude creates rights that are similar to those of a severed mineral interest owner in a common law state.

However, a mineral servitude is subject to *prescription for nonuse*. The Louisiana Mineral Code provides that a mineral servitude will be extinguished by nonuse for 10 years. To interrupt running of the prescription period, operations for discovery and production on the land or property pooled with it are required. Operations need not be successful, but they must be a good faith attempt to discover and produce.

C. LEASEHOLD INTEREST/WORKING INTEREST

The leasehold interest, sometimes referred to as the *working interest* or *operating interest,* is the right to the mineral interest granted by an oil and gas lease. Whether the leasehold interest includes all the incidents of the mineral interest depends upon the precise wording of the granting clause as well as upon the interpretation given by the courts of the various states. However, the lessor typically retains a possibility of reverter of the mineral rights (if the lease terminates), and a royalty interest in production.

D. SURFACE INTEREST

The surface interest is what is left of the bundle of rights of ownership of land after the mineral interest has been severed. Just as the mineral interest is somewhat more than ownership of the minerals themselves, the surface interest's rights are both broader and narrower than to the surface of the soil. The surface interest is more than the right to the surface of the land; it is all rights that are not included in the mineral interest. Therefore, the surface interest has rights to many substances (such as potable ground water) and to many uses (such as use of geologic formations for storage of natural gas) that are not commonly thought of as a part of the surface.

On the other hand, the right of the severed surface owner to the surface is not absolute. As is discussed at pages 159–160, the surface interest's ownership of the surface of the land is subject to the easement of the mineral interest owner or his oil and gas lessee to use as much of the surface when, where and in such ways as is reasonably necessary to search for, develop and produce the minerals. In this respect, the surface owner's right to the surface is encumbered by and servient to the rights of the mineral interest owner.

Disputes between surface owners and mineral interest owners or lessees are common in oil and gas development. A major cause of conflict is a lack of understanding by many surface owners of the nature of their interest. Generally, purchasers of land from which the mineral interest has been severed understand that as surface interest owners they have no right to develop, to lease or to share in the proceeds of leases. However, they often do not understand that their rights are servient to the rights of the mineral interest owner or his lessee, who does not need permission from the surface owner to use the land surface for oil and gas development.

E.   ROYALTY INTEREST

Another commonly encountered oil and gas interest is the royalty interest. A royalty is a share of production free of the costs of produc-

[*37*]

tion, when and if there is oil and gas production on the property. Oil and gas royalties are usually expressed as fractions (e.g. ⅛ of production) or percentages (e.g. sixteen and two-thirds percent of production), but royalty interests for other minerals are often stated as a stipulated amount of money (e.g. $2.00 per long ton).

1. KINDS OF ROYALTY INTERESTS

Several kinds of royalty interests are seen frequently. A *landowner's royalty* is the interest in production retained by the lessor in the royalty clause of the oil and gas lease. It is the mineral interest owner's compensation under the lease after production is obtained. An *overriding royalty* is a royalty interest carved out of the lessee's interest under an oil and gas lease. Overriding royalties are frequently used to compensate landmen, lawyers, geologists or others who have helped to structure a drilling venture. Since an overriding royalty interest is a creature of an oil and gas lease, it ends when the lease from which it is carved terminates. A *non-participating royalty* is a royalty carved out of the mineral interest, entitling its holder to the stated share of production without regard to the terms of any lease. Non-participating royalties are frequently retained by mineral interest owners who sell their rights. A *term royalty* is a royalty carved out of the mineral interest for a stated term, which may be fixed (e.g. for 25 years) or defeasible (e.g. for 25 years and so long thereafter as there is production from the premises). A *perpetual royalty* is a royalty that may extend forever; it is not limited in time. In Louisiana, a *mineral royalty* is similar to a defeasible term royalty in a common law state; it is subject to prescription for nonuse and will terminate in ten years if production does not occur.

2. CHARACTERISTICS OF ROYALTY INTERESTS

All of the various kinds of royalty interests have at least two things in common:

a. *They are not cost-bearing*—By definition, a royalty is free of production costs. However, royalty owners may be required to share costs subsequent to production.

b. *They do not have the incidents of ownership of a mineral interest*—A royalty interest has no right to search for, develop and produce minerals from the property which it burdens and, therefore, no right to use the land surface, except to collect the royalty share. Since the royalty interest has no right to explore, develop or produce on the property, it has no right to grant a lease to another for any of those purposes (except for the landowner's royalty interest, the holder of which retains a possibility of reverter to the mineral interest which it can lease in a "top lease," as is discussed at page 147). Since the royalty interest is not entitled to lease for development, it generally has no right to share in the lease proceeds.

[*39*]

Confusion often arises as to whether an interest created may properly be termed a "royalty" as it has been defined here. The term is often used imprecisely by parties to conveyances, as is discussed at pages 118–122. Disputes arise frequently as to how royalty should be calculated, as well, as is discussed at pages 261–269.

F. PRODUCTION PAYMENT

A production payment is a share of production from the property, free of the costs of production, that terminates when an agreed sum has been paid. Production payments are used in the oil and gas industry for a variety of purposes associated with lease acquisition and financing development, and often take the place of mortgages of producing property. They are similar to a royalty interest that terminates when a specified amount has been paid. An example might be "$^1/_5$ of the oil and gas produced and saved from said land until the market value at the well of such production shall aggregate One Million Dollars ($1,000,000.00)." Of course, it is important to be precise in defining how production is to be valued.

G. NET PROFITS INTEREST

Another oil and gas interest closely related to a royalty interest is the net profits interest. Like a royalty, a net profits interest is expressed as a fraction or percentage of production. Like a royalty, it is non-operating and free of the costs of

[*40*]

production. However, it is different from a royalty interest in that it is payable only if there is a net profit.

The methodology for determining when a net profit has been made is crucial when net profits interests are created. Sometimes "net profits" will be defined so that costs of exploration, drilling and completing are taken into account, as well as operating costs. Or, only operating expenses may be considered. Whatever the meaning intended by the parties, it is important that they define net profits carefully and completely, because reference to a "net profits interest" in and of itself is ambiguous.

Net profits interests are frequently used in addition to or in place of royalty interests as an incentive for a mineral interest owner to grant a lease or as compensation for services. One bargaining for a net profits interest may be able to negotiate a higher percentage net profits interest than he could a royalty interest because the net profits interest will cost the paying party only if there is a profit, while a royalty interest is payable even where expenses exceed revenues.

H. CARRIED INTEREST

A carried interest is a fractional interest, usually created from an oil and gas lease, free of some or all costs. Often, the interest is carried "to the casing point," the point at which the well has been drilled to the desired depth and a decision must be made whether or not to place production pipe, called casing, in the hole and proceed to complete the well for production. If so, it is free only of the costs of drilling and testing preparatory to completion. It is still liable for its share of the costs of completing, equipping and producing the well. In such a case, the carried interest is very much like a working interest except that it is free of the costs of drilling. On the other hand, an interest may be carried "to the tanks or pipeline," which is probably intended to mean that it is free of all costs of completing and equipping the well, as well as of drilling costs. However, it may be argued that the term means that the interest is free of all costs of operation too. If so, the carried interest is tantamount to a royalty.

To complicate matters even more, a carried interest may be (but is not always) subject to the right of the parties paying the costs attributable to the carried interest to recover those costs or even some multiple of them. A common provision in joint operating agreements is that an owner who does not wish to participate in the drilling of additional wells may elect to be carried for the costs of drilling and completing the additional wells. If he makes that election, however, the agreement provides that he will receive none of the proceeds of the production until the parties who put up the money to "carry" his interest receive some multiple of the costs they have expended with respect to the carried interest; usually the multiple ranges from 100% to 500%.

Use of the term "carried interest," like that of "net profits interest" and "production payment," is more of an art than a science. By its very nature, it is imprecise and capable of infinite variations; the list of costs for which an interest may or may not be carried is a long one. Therefore, reference to a "carried interest" can be relied on only as a general description, and the term must be defined fully in the agreement.

I. OTHER INTERESTS

The kinds of oil and gas interests discussed previously are not exhaustive. Owners of mineral rights are free to create their own hybrids and frequently do. What we have considered are the most commonly seen interests, however. Frequently lawyers and the courts will deal with hybrid forms by relating them to more common interests.

CHAPTER 4

PROTECTION OF OIL AND GAS RIGHTS

As discussed in Chapter 2, the rule of capture protects the developer of oil and gas against liability for drainage from the lands of another. However, the rule of capture does not protect a trespasser, whether the trespass is to the surface of the property or to the subsurface. Where an operator drills upon land to which he does not own the mineral rights or drills at an angle into the subsurface of property upon which he does not have the right to operate, liability will be imposed by the same principles that protect interests in real property.

Mineral interests and leasehold interests are protected against trespass by awarding the owner whose rights are infringed compensation for (A) damage to the lease value of the interest, (B) slander of title, (C) assumpsit, and (D) conversion and ejectment.

A. DAMAGE TO LEASE VALUE

Recovery for damage to the lease value of the owner's property is an application of the tort of interference with prospective advantage. Drilling an oil and gas well is the only sure way of "proving" a property. Drilling a dry hole or a

[44]

poor well may "condemn" a property or a formation for oil and gas development by proving that there are not commercially profitable amounts of oil and gas present. Where that happens as a result of a trespass, the owner of the property may recover from the trespasser the amount of the damage to the lease value of the property.

The classic case illustrating the remedy of damage to the lease value of property is *Humble Oil and Refining Co. v. Kishi*, 276 S.W. 190 (Tex. 1925), on rehearing 291 S.W. 538 (1927). There, Humble held a lease dated December 23, 1919, but signed and acknowledged by its lessor on January 29, 1920. The lease term was for three years with provisions that it could be extended by commencement of drilling operations leading to production. Humble took the position that the lease extended for three years from the date it was signed and acknowledged. Shortly before the end of January, 1923, Humble commenced drilling operations and drilled a dry hole. Subsequently, the court determined that the term of the lease ran for three years from its date, so that the lease had expired before Humble commenced its operations. Damages were awarded to the property owner for the bonus value of his right to lease, although there had been no offer from another purchaser.

1.   RATIONALE OF THE REMEDY

Some argue that there has been no real damage to the owner where a dry hole is drilled by a trespasser because the property was worthless for oil and gas development in the first place. That argument ignores economic realities. When property is leased, the lessee customarily pays the lessor a bonus, a payment for executing the lease. The amount of the payment reflects the potential risks and rewards assessed by the lessee, as well as the competition for leases in the area at the time. The market mechanism takes into account the risk that there will be no oil and gas under the property leased when it sets the bonus price. All property has some economic value for lease purposes, and there is a "real" loss to the true owner when a trespasser condemns the property by his actions.

2. MEASURE OF DAMAGES

An interesting question is how damages should be measured—at what point in time is the lease value of the property determined? In order to understand the issue, it is helpful to note that the value of property for leasing generally increases as drilling progresses, at least until there are indications that a dry hole will result. Should damages for destruction of the lease value of the property by the trespasser be measured by the value of the property when the trespass begins or by peak value just before it becomes apparent that no oil and gas will be found? The answer should be that damages will be the difference between the peak value of the property and its value after condemnation, because trespass is a continuing tortious act. Thus, the amount of the potential liability is sizable.

B. SLANDER OF TITLE

Trespassers to oil and gas interests may also be held liable in tort for slander of the owner's title. Slander of title is malicious publication of false statements that are injurious to the plaintiff's title to property or to its quality. Generally, the elements of proof are viewed as (1) a false claim of title, (2) asserted with malicious intent, (3) that causes pecuniary damage. Where these elements of proof are met, the trespasser may be held liable for the amount of the damage suffered by the true owner.

1. FALSE CLAIM

The first element, that the owner show that there has been publication of a false claim to the property, may be met merely by showing that the trespasser occupies the property. It will be met also by a showing that an oil and gas lease that purports to cover the owner's interest in the premises has been recorded.

2. MALICIOUS INTENT

To prove malice, the owner does not necessarily have to show that the wrongdoer acted with evil intent. All that is necessary is to show that the slander was deliberate conduct without reasonable cause. Reasonable cause will be found where the slanderer had a good faith belief in the superiority of his own claims, particularly where he is acting upon advice of his lawyer. However, a good faith belief will not protect against liability for recklessness, and there is a tendency for the courts to be restrictive in defining good faith. Frequently, the improbability of the slanderer's assertions that he acted in good faith is used to infer malice.

3.   SPECIFIC DAMAGES

In contrast to the remedy for damage to the lease value, slander of title requires a showing of actual loss. Proof of specific damage for slander of title to a mineral property usually consists of a showing of a loss of contract or opportunity to sell or lease. The plaintiff must provide the names of those who have refused to deal with him because of the cloud on his title or explain why it is impossible to do so.

C.   ASSUMPSIT

Assumpsit is an equitable action brought to enforce an implied contract. In the context of a trespass to oil and gas interests, it is a suit to enforce payment for the right of entry that should have been obtained. Where the trespass has been in the course of a geophysical search, the suit in assumpsit will claim the value of the geophysical permit covering the property. Where the trespass has taken the form of actual drilling, a suit in assumpsit will claim the lease bonus that should have been paid.

Where the trespass was made in reliance upon a grant of the right from another and that grant contained a warranty of the right, the trespasser theoretically should be able to recover damages he might have to pay the true owner from the person who improperly granted the right. However, as a practical matter, oil companies refrain from suits for breach of warranty for fear that they will discourage other mineral owners from dealing with them. Furthermore, recovery for breach of warranty is generally limited to the amount of compensation paid, plus interest, and that may be less than the damages awarded to the true owner who sues in assumpsit.

D. CONVERSION AND EJECTMENT

A final theory upon which an owner may assert a claim for relief against a trespasser is ejectment and conversion. Such a suit will demand that the trespasser be removed from the premises and be required to account to the true owner for the personal property (the production) sold.

1. BAD FAITH TRESPASS

As is the case with trespass against real property generally, if the trespasser is found to have been acting in bad faith, he is permitted no set off for expenses incurred or benefits conferred. His improvements upon the property and all income from them belong to the owner. Further-more, unless the owner demands it, the trespass-er will not be permitted to plug and abandon a well capable of commercial production; that would be waste.

2.   GOOD FAITH TRESPASS

If the trespasser is found to have committed his transgression in good faith, he will be entitled to recover from production his actual costs or their reasonable value, whichever is less. Thus, if the trespasser exercised superior business and technical judgment and obtained a producing well at a rock bottom price, he will be permitted to re-cover only his actual expenditures. But if, with the benefit of hindsight the trier of fact deter-mines that expenditures were not wisely made, the good faith trespasser will be permitted to re-cover only that portion of the cost deemed pru-dent. The effect of the rule is to remove all pos-sibility of economic benefit for the trespasser.

In considering whether costs incurred by good faith trespassers are reasonable, the courts have generally discussed whether the expenditures for which reimbursement is sought have been of ben-efit to the true owner. The rationale of the anal-ysis is that even a good faith trespasser should be able to recover from the owner only those ex-penditures which would otherwise unjustly enrich the owner.

The "benefit" test is difficult to apply. In the broad sense of the word, even a dry hole is of benefit to the true owner; it will, at the very least, show where *not* to drill. Because of the uncertainty of the benefit test and because of a perceived policy that oil and gas development should be encouraged, most commentators and some courts have suggested that the good faith trespasser should be able to offset all expenditures incurred in exercise of good faith business judgment.

E. CONCLUSION

Because of the nature of the remedies discussed, damage to lease value, slander of title and assumpsit are claims that are generally asserted when the trespass has effectively destroyed the economic value of the property. Where the trespasser discovers oil or gas in commercial quantities, particularly where oil and gas have been produced for a substantial period in large quantities, conversion and ejectment are likely to be sought. In the appropriate circumstances, any of the remedies discussed may impose a heavy burden on a trespasser. They provide an important negative incentive to respect the rights of others.

PART II

CONVEYING OIL AND GAS RIGHTS

CHAPTER 5

CREATION AND TRANSFER OF OIL AND GAS INTERESTS

Oil and gas interests may be created or transferred by conveyance, inheritance, judicial action, or adverse possession. In this chapter we will analyze the basic principles of creating and transferring oil and gas interests.

A. BY CONVEYANCE

A *conveyance* is a transfer of ownership by a presently operative instrument intended to pass ownership of an interest in land to a transferee. Oil and gas conveyances are usually subject to the same formalities as real property conveyances. There are five formalities commonly required. They are (1) a writing, (2) words of grant, (3) an adequate description, (4) designation of the parties grantor and grantee, and (5) proper execution.

1. WRITING

The Statute of Frauds seeks to avoid fraud and perjury with respect to real property (and certain contracts) by requiring that there be a writing signed by the party to be charged with the interest created. Oil and gas interests are treated like real property, under the Statute of Frauds. They must be created and conveyed in writing.

The Statute of Frauds will be satisfied even by an informal writing such as a letter, but most oil and gas interests are created and transferred by formal recordable legal documents entitled "deed" or "lease". Oil and gas conveyances look very much like their real property counterparts. Several sample conveyances are included in the Appendix.

a. Deeds

There are two general types of deeds in use in the United States to convey oil and gas interests. The basic distinction between them is the presence of covenants or warranties of title.

(1) *Warranty Deed*

A warranty deed grants the property described with covenants (promises) of the grantor as to title. A warranty deed may contain up to six overlapping covenants by the grantor as to title: seisin, right to convey, no encumbrances, warranty, quiet enjoyment and further assurances. Covenants obligate the grantor to protect the grantee and those who take from him against conflicting claims to the interest granted, to the extent of returning the money paid to the grantor, plus interest.

Warranties may be specified in deeds or, in some states, may be incorporated by reference (e.g., "I grant with general warranty covenants . . . .") or implied from the use of certain granting language (e.g., "I grant, bargain, sell and convey the following described land . . . ."). A warranty deed that contains all six covenants of title is sometimes called a "full" or "general" warranty deed. A deed is also sometimes called a "general" warranty deed if it includes a promise to protect the grantee against the claims of "all persons whatsoever" or similar language. A warranty deed that contains fewer than all of the covenants of title or that limits the scope of the warranties to protection against persons claiming "by, through, or under the grantor or his heirs" may be called a "special" or "limited" warranty deed.

(2) Quitclaim Deed

A quitclaim deed contains no covenants of title. The grantor grants whatever interest he may have to the grantee, but without any guarantee that he has any interest to grant. Whatever rights the grantor has to the property described, he "quits" or releases to the grantee.

Deeds without covenants of title will usually be titled "Quitclaim" and use that word in the granting clause (e.g., The grantor quitclaims and conveys the following described property . . .). However, in many states any deed without an express statement of covenants of title is a quitclaim deed.

(3) Importance of Title Covenants

Title covenants or warranties serve two important functions in oil and gas conveyancing:

(1) if the covenants are breached, the grantee is entitled to recover damages suffered up to the amount of the consideration paid, plus interest and expenses incurred in defending the title; and

(2) the presence of the covenants gives the grantee the protection of the doctrine of estoppel by deed, which may pass after-acquired title of the grantor.

For these reasons, most transactions creating or transferring oil and gas rights are completed with deeds containing covenants of warranty.

However, there are frequent exceptions. Transactions between persons active in the oil and gas industry are often completed on specially drafted forms without warranties of title or on printed forms with the covenants struck out, particularly where the grantor's compensation is in the form of a retained interest. Quitclaim deeds are used as a matter of course to clear clouds on title.

b.   Oil and Gas Leases

Oil and gas leases are usually granted on printed forms, as are deeds of mineral and royalty interests. There are many commercial printing houses that offer a wide variety of lease forms. Two typical examples of oil and gas lease forms are included in the Appendix. Bear in mind, however, that there are wide variations in forms from state to state and even within states. In addition, many oil companies, lease brokers and some large mineral interest owners have drafted and printed their own lease forms.

Oil and gas leases are different from ordinary leases of real property in at least three respects:

(1) the lessee acquires not only the right to use the premises but also the right to take substances—the oil and gas produced—from the land;

(2) the lessee's rights do not necessarily end after a term of years. In fact, they may be perpetual; "as long as oil and gas is produced;"

(3) the lessee's right to use the land is not exclusive; it is subject to the surface owner's uses that do not interfere with the lessee's efforts to acquire the substances covered by the lease.

For these reasons, oil and gas leases are generally treated by the courts like deeds of easement, or deeds creating a *profit a prendre*, or even deeds to the minerals in place, rather than leases of real property.

Oil and gas leases generally contain covenants of title for the same reasons that they are included in mineral and royalty deeds—to give the grantee some protection against defects of title and the benefit of the doctrine of after-acquired title. However, most printed forms contain only the covenant of warranty, obligating the lessor to protect the lessee against actual or constructive eviction by one with paramount title. As competition for leases has increased, it has become common for all title covenants to be deleted or disclaimed by lessors.

c.   Other Instruments

Other instruments commonly used in the creation and conveyancing of oil and gas interests include assignments of interests, grants of right of way, mortgages, deeds of trust and numerous documents closely related to real property conveyancing. As a general rule, the adequacy and effect of these is governed by the same principles of law that control real property conveyancing.

2.   WORDS OF GRANT

An instrument purporting to convey an interest in oil and gas, like other interests in real property, must contain words of grant. No "magic" language is required; it is sufficient that the language be clear enough to show the grantor's intention that there be a present transfer of a present or future interest. A statement that "I give" or "I transfer" the interest would probably suffice. Obviously, the good draftsman will not leave the matter open to question. On the premise that more is better, most deeds and leases contain detailed words of grant (e.g., "grant, bargain, sell, convey, transfer, assign and deliver").

3.   DESCRIPTION

The third formality for the creation or transfer of oil and gas interests is that there must be an adequate description of the property to which the interest attaches. In practice, this requirement divides into two standards, (1) legal validity and (2) marketability.

a.   Legal Validity

The standard of legal validity must be met for the instrument to be effective between the grantor and the grantee. It is not a high standard. The courts apply the same rule as is applied to real property interests in general and hold the description legally adequate if it is sufficient to permit location of property with reasonable certainty. A description may be legally valid even if oral or other extrinsic evidence is necessary to locate the property; e.g., a grant of "$1/32$ perpetual non-participating royalty in the 40 acres upon which the house that Uncle Charlie built for Aunt Mary sits" might well be legally valid despite the unorthodox description.

b.   Marketability

The marketability standard requires a description sufficiently certain to make the title freely assignable in commerce. The standard for marketability is higher than that for legal validity. Though reference to the property where "the house that Uncle Charlie built for Aunt Mary sits" might be legally valid, it clearly would not be sufficient for a New York, Denver, Houston, or Los Angeles banker to lend money upon. The marketability standard requires location of the tract solely by reference to the public records, without ambiguity, uncertainty, or reference to extrinsic facts.

c.   Methods of Description

Most descriptions in conveyances of oil and gas interests meet both the standard of legal validity and the standard of marketability. Generally, one of two description systems, or a combination of the two, is used in oil and gas conveyances as in ordinary real property transactions.

(1) *Reference to Government Survey*

The more frequently used description system is to locate the property by reference to government survey, identifying it in the terms of one of the many land surveys conducted under governmental authority. The most common government survey description is reference to the "standard" or rectangular system established by Congressional fiat in 1785. Under the "standard" system, there are established six mile square "townships", located by reference to imaginary lines running north and south (principal meridians) and east and west (principal base lines). Each township is composed of 36 one mile square "sections" of approximately 640 acres each.

Use of the "standard" system results in references to townships, sections and quarter sections; e.g., "the Southeast Quarter of the Southwest Quarter of Section 13, Township 12 North, Range 5 West of the Indian Meridian, Canadian County, Oklahoma." It is important to note, however, that the "standard" system is used only in approximately half the states—and not exclusively in many of those. In Texas, for example, there have been surveys under four different governments—the Spanish, the Mexican, the Republic of Texas and the State of Texas—none of which use the "standard" system.

(2) Metes and Bounds

Where property is not described by reference to government survey, it is usually described by metes and bounds. A metes and bounds description locates property by reference to its exterior boundary lines. It is expressed in terms of natural or artificial "monuments" (such as creeks, rocks, and stakes) and directions and distances. Metes and bounds descriptions tend to be lengthy and poetic; e.g., "beginning at the granite boulder on the north side of the bridge over Oil Creek, thence Northeasterly 280° thirty rods to an iron stake, thence East by Northeast to the white oak tree . . . ."

One problem with metes and bounds descriptions is that it may prove difficult to locate the monuments. The white oak tree and the granite boulder referred to above may have looked distinctive to the surveyor who drew the description, but time pulverizes even granite boulders, and oak trees multiply, albeit slowly. Another problem is that the length of metes and bounds descriptions increases the risk of errors in copying from instrument to instrument. If the error is in a direction, the description may not "close" (i.e., the lines of the boundary may never meet). If the error is in a distance, there will be a gap in the boundary. Elaborate rules for rationalizing ambiguities or errors in description have been developed by the courts.

One special application of the metes and bounds description method is the recording of plat maps that permit incorporation of metes and bounds descriptions in deeds by reference to lots and subdivisions. Another is the "bounded by" method sometimes used in oil and gas leases when it is inconvenient for the person taking the lease to obtain a full legal description. In such circumstances, particularly in the Appalachian states, oil and gas leases may describe the property leased by reference to the ownership of surrounding properties at the time the lease is granted; for example: "Bounded on the North by the lands of John Schur, on the East by lands of Harry Sauer and lands of Sarah Staley Lowe, on the South by the lands of Florence and John Lowe and on the West by State Route 161 and the lands of Floyd Moine." "Bounded by" descriptions may appear strange to those not used to them. They may be confusing where ownership of surrounding properties has changed between the time of the grant of lease and the attempt to locate the property. However, "bounded by" descriptions are legally valid; it is possible to locate the property. But the marketability of such descriptions may be questionable, and a survey or metes and bounds description is preferable.

4. PARTIES DESIGNATED

a. Identification of the Parties

There are two aspects of the requirement that a conveyance identify the grantor and grantee. The first is that an instrument must identify the parties grantor and grantee with reasonable certainty. The rationale of this rule is certainty and concern that seisin, the magic substance of property ownership, must always rest in someone. Compliance with this formality is usually a matter of making sure that all of the blanks of the deed form are completed.

b. Capacity of the Parties

The second, and more troublesome, aspect of the requirement that instruments of conveyance designate the parties is that those designated must have *capacity* to be a party. Not everyone has the legal right to make conveyances or to hold property rights. Minors, incompetents and drunkards, for example, all lack or have limited capacity to transfer interests in land. In oil and gas conveyancing, common capacity problems involve Indians, attorneys in fact, married couples, and concurrent and successive owners. Some aspects of acquiring interests from persons with limited capacity are discussed in Chapter 6.

5. EXECUTION

Execution, as that term is used with reference to conveyances in general, means completion of the instrument. Execution may involve as many as four separate elements: a) signature, b) attestation and acknowledgment, c) delivery and acceptance, and d) recording.

a. Signature

Instruments transferring oil and gas interests are required by the Statute of Frauds to be signed, since they create interests in land. Signature of a deed or lease attests to its validity. It is usual to affix one's business signature rather than one's full given name. President James Earl Carter, Jr. signed his name as Jimmy Carter to documents of state. Even an "X" may be a valid signature if it is intended by the person signing it to indicate the validity of the instrument and if other special requirements are met.

Though it is not required that one sign his given name, as a practical matter, it is important that a grantor's signature to a deed or a lease be the identical name shown on the instrument conveying his rights to him. Thus, if the record shows a deed granting the mineral rights in Blackacre to John Taft Lowe, a deed or lease from him at a later date should be signed the same way, and not as John Lowe or John T. Lowe, to avoid any possible question of the chain of title.

Oil and gas instruments are usually in the form of *deed polls;* i.e., they are structured to be signed only by the grantor. Upon acceptance, they are fully as binding upon the grantee as instruments in the form of contracts signed by both parties. However, it does no harm (and may be advisable to put the grantor at ease where the instrument contains promises from the grantee) to have the grantee sign the instrument as well.

Consideration is not generally required to support creation or transfer of oil and gas interests. As with conveyances of other real property rights, title passes if the instrument is properly executed and delivered. However, most oil and gas conveyances contain recitals of consideration to avoid creation of a resulting trust and to qualify the grantee as a bona fide purchaser for value under applicable recording statutes.

In fact, consideration is bargained for in most oil and gas transactions, and if it is not actually paid there may be grounds for the grantor to rescind the deed or lease and recover title. Furthermore, in a minority of states, there is precedent that oil and gas leases are contracts and must be supported by consideration. Louisiana goes even further and requires "serious" consideration.

b. Attestation and Acknowledgment

Attestation, or witnessing, is the practice of having persons who are not parties to an instrument testify that they saw the grantor sign the instrument (or, sometimes, that they recognize his signature) by affixing their signatures to the document as witnesses. *Acknowledgment* is affirmation under oath by the grantor that the signature is his own and, usually, that he has the authority to sign and does so freely. Acknowledgments are usually given before a notary public, but in many states, recording clerks, judges, lawyers or other officials are empowered to take acknowledgments as well.

Generally, neither attestation nor acknowledgment is necessary for a valid conveyance. If an instrument complies with all other formalities it is valid as between the grantor and the grantee without attestation or acknowledgment or with improper attestation or acknowledgment. However, joinder of the spouse or special acknowledgment may be required to validate a conveyance of property subject to marital rights. Moreover, in most states, proper attestation, or acknowledgment, or both are necessary to qualify a conveyance for recording.

Improper attestation or acknowledgment occurs frequently in oil and gas transactions. Probably the most common defect is that the grantor is not presented personally before the oath-giving officer to acknowledge the instrument, as most states require. For example, the landman who acquires the lease may take it back to his office in the city for acknowledgment. Another common defect is witnessing or acknowledgment by an employee or agent of the grantee whose action may be challenged on the ground that he had an interest in the transaction.

Where a defective attestation or acknowledgment is challenged, usually by a subsequent purchaser who seeks to avoid having notice imputed by the recording statutes, the states split into three groups:

1. The strictest position is that the defectively attested or acknowledged instrument should not have been allowed on the record and, therefore, will be ignored. The defect makes the recording ineffective to give notice even to those persons who may actually have

seen it on the record. Recording gives no constructive, actual or inquiry notice;

2. The most liberal position, adopted in Colorado by statute, is that if the defectively attested or acknowledged instrument is placed on the record it serves as constructive notice to the whole world. Though it should not have been recorded, its defects will be ignored if it is actually recorded;

3. An intermediate position is that a defectively attested or acknowledged instrument does not give constructive notice (since it should not be on the record), but it may put those who see it on actual or inquiry notice of the grantee's claim to an interest.

Potential problems with defective attestation and acknowledgment of oil and gas interests have become real only infrequently in the past. However, oil and gas interests are worth much more today than only a few years ago, so we may expect to see more litigation over such issues.

c. Delivery and Acceptance

The third element of execution is delivery and acceptance. *Delivery* is any act that shows clearly the intent of the grantor that title be passed presently. Usually, delivery takes place when the deed is handed over, but physical transfer of the instrument is neither required nor conclusive that delivery has taken place.

Delivery turns on the facts. It may have occurred though the instrument is in the hands of a third party or even though the grantor still has it. Or, there may be no delivery even though the deed or lease has actually been given to the grantee. What the courts look for in each case are facts indicating the intent of the grantor and the grantee to pass title presently, without conditions precedent or right of recall.

Acceptance is a showing by the grantee that he wishes the transfer to be effective. With conveyances of oil and gas interests as with real property transfers generally, acceptance is usually implied from the fact that the grantee takes the instrument; the grantee does not usually sign the deed or lease.

Disputes over delivery and acceptance are probably more common in oil and gas conveyancing than in real property conveyancing generally. This is because oil and gas interests are usually created and transferred without formal "closings", gatherings at which the deed or lease is signed, witnessed and acknowledged and the agreed consideration paid. Often, the transaction is conducted by mail and over the telephone.

As a practical matter, the moral for grantees is "Get the deed or lease in hand". There is a strong presumption that an instrument in the possession of its grantee has been delivered and accepted.

[*68*]

d.   Recording

The final step in conveying oil and gas rights is recording. In most states, recording is a practical requirement for validity rather than a legal requirement. As a general rule, an instrument is valid as between the grantor and the grantee even though it is not recorded. Recording protects the grantee against claims of subsequent purchasers or creditors.

However, recording may be a legal requirement. In Kansas, recording or registration for taxation within a specified time is required to validate a mineral deed. In several states that have enacted marketable title acts or dormant minerals acts (see the discussion at page 89), severed mineral interests can be preserved beyond the statutory limitations period only by special recording or use.

B.   BY INHERITANCE

Oil and gas interests may be acquired by inheritance, as a result of the provisions of a will or of the intestacy laws, as well as by conveyance. Where inheritance is the basis for the creation or transfer of such interests, the requirements that have to be met are those that apply to probate law and to estates generally. No problems peculiar to oil and gas interests are created.

C. BY JUDICIAL TRANSFER

Oil and gas rights may also be transferred by judicial action. That occurs where there is foreclosure of a mortgage or some other lien encumbering the property. Tax sales and administrators' or executors' sales are other examples. So too is the order of a conservation agency compulsorily pooling property.

The requirements for valid creation or transfer of oil and gas interests by judicial action are strict and a substantial source of litigation. Minor deviations from the statutory procedures will be considered to be "mere irregularities" that will not invalidate the transfer, but more serious "jurisdictional defects" will. To avoid jurisdictional defects, the court that orders the transfer must have proper jurisdiction of the subject matter (including the amount), the property, and the parties to the litigation.

The intricacies of jurisdiction of the courts is beyond the scope of these materials. As with transfer of interests by will or intestacy, the issues presented are not peculiar to oil and gas law.

D. BY ADVERSE POSSESSION

It is also possible to acquire title to oil and gas interests, like other real property interests, by using them like an owner. Despite the strong interest in permitting the public to rely upon record

[*70*]

title, where one adversely possesses property by using it "like an owner" for a sufficient period of time, fairness and economic efficiency demand that he be recognized and legally protected as the owner.

Generally, what is required to establish adverse possession is possession of real property in an open and visible manner, continuously and exclusively for the limitation period, under a claim of ownership sufficient to put other parties on notice that the adverse possessor claims as an owner. Often, adverse possession will be under *color of title*, under a written instrument that the adverse possessor believes conveyed the property to him. However, color of title is not necessary for adverse possession, except in a few states. It is not necessary that the possessor personally hold the property for the full limitation period; where there is privity between possessors, their time in possession may be "tacked" together to meet the requisite period.

In modern times, the doctrine of adverse possession has been codified in statutes of limitations. Where the requisites of adverse possession are met for the statutory period, which ranges from five to twenty-five years, the record owner is barred by statute from suing to eject the adverse possessor or to quiet title. The adverse possessor may be entitled to a decree establishing a new and original title to the premises.

Though transfer of oil and gas interests by adverse possession is probably less common than transfer by heirship or judicial sale, application of adverse possession to oil and gas interests creates special problems. These are solved by application of fundamental principles.

1. ARE BOTH THE SURFACE AND MINERALS ADVERSELY POSSESSED?

A common problem of adverse possession of mineral properties is the scope of the possession; does the adverse possession extend to both the surface and the minerals? That issue is answered by applying of the principles of unity of possession, relation back, and paper transactions.

a. Unity of Possession

A fundamental principle of adverse possession is that the adverse possessor takes all that the record owner against whom he adversely possesses has. Therefore if:

O owns fee simple absolute; and

A adversely possesses *by farming* the surface for the statutory period;

A acquires title to both the surface and the minerals when the statutory period ends. But if ownership of the minerals has been severed from the surface when the adverse possession begins, so that:

O owns the surface interest only; and

X owns the mineral rights; and

A adversely possesses by farming the surface
for the statutory period;

A acquires title only to the surface rights.

The courts distinguish the situations on the basis of to whom notice is given by A's adverse possession. Where O owns the fee simple absolute,
O knows or ought to know that adverse possession of surface by farming is a claim by A to the
mineral interest as well as the surface interest.
However, where the mineral interest has been
severed from the surface, possession of the surface gives no notice to the severed mineral interest owner because surface use is not inconsistent
with the rights of the mineral owner. Another
way of rationalizing the result is by the public
policy in favor of unity of title; where there is
ambiguity the courts rule in favor of less title
fragmentation because the public has an interest
in efficient use of property, and that is more likely to follow where there are fewer (rather than
more) owners of interests.

b. Relation Back

A second fundamental principle is that title
earned by adverse possession relates back to the
time of its beginning. It is not affected by severance of the minerals from the surface after adverse possession has begun. Therefore, if

O owns fee simple absolute; and

A begins adverse possession by farming the
surface; and

O then severs the minerals by conveying them to X;

A gains title to the fee simple absolute when the statutory period runs. A's title relates back to the beginning of adverse possession. Another way to understand the result is to see it as an application of the principle that one can give no better title than he has; O could only give the mineral rights subject to the claims of A.

c.   Paper Transactions

Why does the conveyance from O to X in the last example not interrupt A's adverse possession? Because once A takes possession of the property adversely, he must be physically or constructively dispossessed to interrupt the adverse possession. A mere paper transaction is not enough. However, it would be sufficient to interrupt adverse possession if X commenced drilling operations.

2.   WHAT MUST BE DONE TO ADVERSELY POSSESS SEVERED MINERALS?

Where mineral rights have been severed from the surface interest, mere use of the surface for the statutory period will not be sufficient to establish title by adverse possession. However, except in Louisiana, title to severed minerals can be acquired by actually taking the minerals for the statutory period.

The cases generally say that title by adverse possession to severed minerals requires a continuous taking of the minerals for the statutory period. If such statements are taken literally, adverse possession of severed minerals will not begin until actual production is obtained from the land. The courts and commentators have generally concluded that notice of the adverse claim is the key to adverse possession. If so, actions less than actual production of the minerals should constitute adverse possession. If notice is the issue, adverse possession should begin when operations for drilling or mining are commenced upon the property. Likewise, the requirement that the adverse possession be continuous for the statutory period should be met by intermittent but obviously unconcluded operations. For example, suppose that A, an adverse possessor, commences drilling operations on January 1 and concludes drilling operations with a dry hole on April 1. He does not restore the access roads or drill pits and leaves pipe and equipment on the property until October 1, when drilling operations for a second well are commenced. A's adverse possession should be held to be continuous from January 1. Although A did not work on the land continuously, his use of it was obvious for all to see and consistent with a claim to the mineral rights.

3. UNRESOLVED ISSUES

Virtually every jurisdiction has extensive precedents on adverse possession. Surprisingly, there are several unresolved issues.

a.　What It Takes to "Sever" Minerals

As has been noted, whether or not the mineral interest has been severed from the surface interest at the time adverse possession begins determines what actions are required to constitute adverse possession. Therefore, it is important to determine what constitutes a severance of the minerals.

There is no doubt that grant of an interest in the minerals by a mineral deed constitutes a severance. But what if the grant is not by a mineral deed but by an oil and gas lease? In states such as Texas that hold that an oil and gas lease conveys an estate in the oil and gas to the lessee, it is clear that an oil and gas lease severs the minerals from the surface. However, in many states, an oil and gas lease is held to give the lessee something less than an estate in the oil and gas— a profit a prendre, a profit in gross, or a license. By the logic of the common law, it is doubtful that such interests would sever the minerals from the surface. Professors Howard Williams and Charles Meyers argue that an oil and gas lease should be treated as a severance of the minerals from the surface for purposes of adverse possession because the rights of entry and use given by an oil and gas lease are substantially identical to those of a mineral deed.

b. How Much of the Mineral Is Acquired

A second unresolved issue of adverse possession of severed minerals is the amount of the mineral earned by adverse possession. For example, if:

O owns the severed mineral rights under a 640 acre section; and,

A adversely possesses for the statutory period by producing oil and gas from the southeast 160 acres;

how much oil and gas does A earn? Does he become owner only of the oil and gas that will be drained by the well he has produced? Or, does he earn title to all of the oil and gas under the spacing unit, which may cover an area larger than that actually drained? Another possibility is that he acquires title to all of the oil and gas under the entire 640 acre tract. Finally, what about oil and gas in deeper (or shallower) formations not being produced by the adverse possessor's well? Are they earned?

The cases give little guidance. In dealing with hard minerals, many courts have said that the adverse possessor earns title only to that amount of the minerals produced or loosened by the mining activities. But there is no reason for a record owner who learns of drilling operations on a portion of a tract by an adverse possessor to conclude that the adverse possessor's claims are limited to that portion. In the interest of unity of title, adverse possession of a part of a reasonably sized tract should give the adverse possessor title to the oil and gas to all depths under the whole spacing unit (if he possesses without color of title) or under the area covered by his instrument (if he possesses under color of title).

c.   What Minerals Are Earned

A parallel problem is whether acquisition to title of oil and gas by adverse possession acquires for the adverse possessor rights to other minerals, such as coal. Some cases have held that adverse possession of hard minerals does not earn title to oil and gas. Again, in the interest of unity of title, the better result would be that adverse possession of one mineral earns title to all minerals belonging to the person against whom there has been adverse possession, under the area worked or under the entire tract, depending upon whether the possession was under color of title.

CHAPTER 6

JOINT OWNERSHIP OF OIL AND GAS RIGHTS

It is common in the United States for property rights to be owned jointly, by more than one person. This is particularly true of mineral rights; fractionalized interests are the rule rather than the exception. Therefore, it is relevant to consider the nature of the rights of joint owners. What kinds of relationships create joint ownership rights? Can one who owns a fraction of the mineral interest grant a lease or develop without permission of the other owners? Whose permission to develop must be obtained and how should it be accomplished?

A. CONCURRENT OWNERS

At common law, and in most states today, there are three types of concurrent ownership:

Tenancy in Common—the joint owners have separate but undivided interests in the property; i.e. each owns a separate fraction, but it is not possible to identify which part belongs to any tenant. Mineral interests are frequently divided into minute fractions in this manner.

Joint Tenancy—each joint owner owns the whole thing, subject to the right of survivorship of the other owners. The last owner alive takes all the interest. A joint tenancy interest may be "severed" by conveyance, which destroys the right of survivorship and converts it to a tenancy in common; e.g. if A, B and C are joint tenants and C conveys to D, D holds his interest as a tenant in common with the joint tenancy of A and B.

Tenancy by the Entirety—this is a form of concurrent ownership available only to husbands and wives. It is similar in effect to a joint tenancy in that each spouse's right is subject to survivorship, but different in concept in that the spouses are treated as one person; they have "unity of person," in addition to unities of time, title, interest and possession. A tenancy by the entirety may not be severed by a conveyance by one spouse, but it will be converted to a tenancy in common or a joint tenancy by divorce.

The common element of all three ownerships is that all of the co-owners have the right to present possession of the property at the same time; their ownership is *concurrent*.

1. DEVELOPMENT BY CONCURRENT OWNERS

The most common problem with concurrent ownership is whether one or more of the owners have the right to develop minerals, or to lease for their development without the consent of the other owners.

Suppose that A and B are tenants in common of Blackacre in fee simple absolute. A owns a 90% undivided interest and B owns a 10% undivided interest. A wishes to develop. A cannot locate B, though he searches diligently (or, B is located but refuses to cooperate in drilling). A proceeds anyway and completes a prolifically producing well. What rights has B?

a.   Minority Rule

In a minority of states, including Louisiana, Illinois and West Virginia, there is precedent that it ·is waste for a cotenant (or a cotenant's lessee) to drill for oil and gas without the consent of the other owners. The rationale of this position is the traditional view that any action that changes the nature or character of jointly owned land is waste, even if it improves it. In such states, A, the cotenant who wishes to drill, may be enjoined from development or held liable for damages as a trespasser, unless he can show that development was necessary to protect against drainage.

b.   Majority Rule

The example is based on the landmark case of *Prairie Oil and Gas Co. v. Allen*, 2 F.2d 566 (8th Cir.1924). There, the 90% tenant in common of the mineral interest leased its interest to an oil company. After production was obtained, Lizzie Allen, the owner of the surface and the remaining 10% mineral interest, sued the purchaser of production and the lessee. She contended that she was entitled to one-tenth of all production from the land, arguing that the oil company had no right to develop without her permission, that it was a trespasser to her interest.

The court rejected Lizzie Allen's argument that development without her permission was trespass. It said that any tenant in common (or his lessee) has the right to remove minerals from the jointly owned property because an interest in minerals can only be enjoyed by developing them. Development is use of the interest, not destruction of it. On that basis, the court required an accounting to Lizzie Allen for her share of the production less her proportionate share of the costs of operating, after all drilling and completion costs had been recovered. It also noted that Lizzie Allen would have had no liability if the well had been a dry hole or had never produced enough to permit the operator to recover his costs.

Thus, the majority rule, adopted in Alabama, California, Florida, Georgia, Kansas, Kentucky, Missouri, Montana, North Dakota, Oklahoma, Pennsylvania and Texas, is that a tenant in common (or his lessee) has the right to develop minerals without the permission of other cotenants, or even over their objection. The developing party must pay all costs, but has the right to recoup costs paid from production. Thereafter, he must account to the nonconsenting owners. Furthermore, he must be careful not to deny the nonconsenting owners' rights to develop independently or to lease for development, for that would be an "ouster" of the other cotenants that would make him liable as a trespasser.

2.   A CRITICAL EVALUATION OF THE MAJORITY RULE

Several notes are in order concerning the majority rule for development by cotenants. First, the rule of *Prairie Oil v. Allen* has *not* been specifically adopted in several jurisdictions that now produce substantial amounts of oil and gas. It is generally regarded as the better rule, because it is closer attuned to the trend of the law of waste and to the effect of oil and gas development. However, a strong argument can be made that nonconsenting cotenants should be able to insist that their share of the production be left in the ground, if not that they should be able to bar development altogether; if fast rising oil and gas prices are to be anticipated, early development is not necessarily in the interest of the owners.

Second, it should be noted that *Prairie Oil v. Allen* involved a tenancy in common, not a joint tenancy or tenancy by the entirety. Its principle should apply equally to a joint tenancy because a joint tenant could convert his interest to a tenancy in common by conveyance. Its principle may not apply to a tenancy by the entirety because tenants by the entirety share "unity of the person"; i.e., the two are one legal entity and may

[*83*]

not act separately. The rights of married persons are discussed at pages 90–92.

Third, the majority rule is not often relied upon. One reason is that there are frequent disagreements over what costs may be recouped by the developing owner. In the context of the example, suppose that the first well on the premises had been a dry hole and that production had been obtained only by drilling a second well. Should the costs of the dry hole be recoverable from the production of the second well? Most of the cases that have considered the issue base their decisions on whether the dry hole was of benefit to the nonconsenting owner. However, the concept of "benefit" has proved to be elusive. A strong argument can be made that the nonconsenting co-tenant should pay his share of all costs that are not unreasonable or incurred in bad faith before sharing in production.

Another and perhaps more important reason for not relying on the majority rule of the rights of concurrent owners is that it confers legal rights that often make little economic sense. Suppose that in our example B has a 50% undivided interest (rather than 10%). If A relies upon the rule in *Prairie Oil & Gas Co. v. Allen*, A will bear 100% of the risk of loss of drilling a dry hole but gain only 50% of the right to production if successful. Unless the prospect is superlative, A is likely to decide not to drill without B's consent because the probable return on investment will not be worth the risk.

3. OTHER METHODS OF OBTAINING THE RIGHT TO DEVELOP

Because of the economic realities, the rule of *Prairie Oil & Gas Co. v. Allen* is not often relied upon except as a last resort or where very small interests are nonconsenting. Instead, statutory or judicial devices are used to obtain the rights of lost or recalcitrant owners.

a. Forced Pooling

The preferred way to obtain nonconsenting interests for development is by forced or compulsory pooling. Forced pooling is the compulsory joinder of ownership rights in property within a proposed well spacing unit by exercise of the state's police power. As is discussed at pages 20–21, it is a legal device developed to permit government to establish minimum sized spacing units without destroying the correlative rights of small tract owners. All but one of the states with petroleum conservation laws have forced pooling sections in their legislation. Approximately two-thirds of these permit application of the forced pooling provisions to undivided fractional interests as well as to separately owned small tracts within the spacing unit.

Forced pooling provisions differ substantially from state to state, but the basic concept is that the state exercises its police power to protect its citizens from over-drilling and the correlative rights owners from drainage by forcing the non-consenting owner to accept administratively determined fair terms. In some states (Oklahoma for one), forced pooling procedures are fast and relatively simple, so that forced pooling is the usual way of dealing with nonconsenting interests. In many other states (including Texas), the forced pooling procedures are so arduous or the scope of the legislation so limited, that forced pooling is rarely used.

b. Judicial Partition

Judicial partition is the division by court order of undivided interests. It was available only to possessory interests at common law. Therefore, it ought not be available to divide concurrent interests in severed minerals in states that have embraced the non-ownership theory of oil and gas rights. However, it has been applied to mineral interests either by statute or by the courts in most states, regardless of ownership theory. Other nonpossessory interests, such as royalty interests or the mineral interest owner's possibility of reverter under a lease, are generally not entitled to partition.

Partition may be *in kind* (a division allocating specific portions to each owner) or *by sale* (conversion of the interests to cash and division of the money). In property law, partition in kind is favored because it disturbs land ownership less.

In a majority of states, partition is a matter of legal right; in those jurisdictions the complaining cotenant is entitled to partition either in kind or by sale, though the courts have the discretion to choose between partition in kind and partition by sale to balance equities. As a general rule, courts will not partition producing or potentially producing property in kind because a division of the land into tracts proportionate in size to the interests of cotenants may not equally divide the minerals. However, there is a strong minority of producing states including Oklahoma and Kansas, in which the courts have the authority to deny partition altogether if necessary to prevent the remedy from being used for "fraud or oppression".

One of the attractions of partition from the view of the partitioning party is that he is usually entitled to recover costs and reasonable attorneys fees from the other owners, on the theory that partition benefits the property. A practical disability of partition is that it requires adversary proceedings in court that may drag on for years. As a result, it is not regarded with favor by the oil industry.

c. Lost Mineral Interests

Forced pooling and judicial partition are remedies for the problem of nonconsenting concurrent owners that may be applied either to recalcitrant mineral interest owners (those who can be located but will not agree to develop or lease) or to lost

[*87*]

mineral interest owners (those that cannot be located). Lost mineral rights are a special problem because of the difficulty of obtaining jurisdiction over a lost mineral interest owner, and some states have taken special steps to deal with them.

In the years just before and after each of the periodic oil booms over the last century, many property owners in oil producing states reserved fractional mineral rights from real estate conveyances hoping that they would become valuable. In many cases, those interests did not become valuable, even for speculation, until the 1970's and 1980's. In the meantime, the severed rights had been further fractionalized by operation of residuary clauses of wills and intestacy laws. They are often owned now by persons who are not aware of their existence. Tracing those persons and purchasing or leasing their rights has become a monumental problem for the oil and gas business—and a growth industry for lawyers and landmen.

In an attempt to unify titles, many states have enacted special legislation to limit the life of severed mineral rights. A detailed consideration of the various statutes is beyond the scope of this book, but the most important statutes can be classified as follows:

1. *Prescription* —As discussed above, in Louisiana mineral servitudes, mineral royalties and leases are extinguished by non-use for 10 years. Tennessee puts a statutory limit of 10

years on oil and gas leases without development.

2. *Marketable Record Title Acts* —Some states have applied marketable record title acts to oil and gas rights so that interests that conflict with a record chain of title are extinguished. If the record does not contain a specific reference to the interest within the statutory period (usually 30 to 40 years), it is destroyed.

3. *Dormant Mineral Acts* —Closely related both to marketable record title acts and to prescription are statutes that declare mineral interests not developed within a stated time, usually 20 years, to be extinguished unless specifically registered. Several state courts struck down such laws, finding that they were in violation of the due process clause of the U.S. Constitution. In *Texaco v. Short,* 454 U.S. 516, 102 S.Ct. 781, 70 L.Ed.2d 738 (1982), the Supreme Court approved the Indiana Dormant Minerals Statute, opening the door to widespread enactment.

4. *Taxation and Sale* —Some states subject severed mineral interests to separate taxation. If the taxes assessed are not paid, the interests are sold at sheriff's sale. A practical problem with such statutes is that officials charged with administering them often neglect to assess taxes because of the expense and time involved. In Colorado, a statute permits the surface owner to require county officials to

assess taxes on the severed mineral interests in his land.

5. *Receivers or Trustees to Lease* —Most major oil producing states have legislation permitting probate courts to appoint receivers or trustees to lease on behalf of lost mineral interest owners upon judicially approved terms. The proceeds from leases are held in escrow and eventually escheat to the state if not claimed. Oklahoma has gone one step further and provided for escheat of the underlying mineral interest at the same time as the proceeds.

B. MARITAL RIGHTS

Both common law and state statutes provide substantial legal protection for spouses against disinheritance. The rights created are a special application of concurrent rights. They also present special problems.

There are three kinds of general marital rights that frequently have an impact on the creation or exercise of oil and gas rights.

1. *Dower* —At common law, dower was the interest a surviving wife received in the inheritable lands owned by her husband during marriage. She received a life estate in one third of such lands. The corresponding interest of the surviving husband was "curtesy," a right given to surviving husbands who had proved their manhood by fathering a male heir born alive, to a life estate in all the property owned by their wives during marriage. In many states the

distinction between the two interests has been abolished by statute so that surviving spouses are entitled to equal interests in property acquired during marriage, (usually a $1/3$ life estate) upon the death of the partner. However, both at common law and in modern times, the record may not show a dower or courtesy interest.

2. *Homestead* —Many states have enacted legislation intended to protect property used as the family home against attachment and sale by creditors. Those statutes function in part by barring creation and transfer of rights to such property unless both spouses join. In some states, homestead must be noted on the deed and it is not required that there be actual occupancy of the land claimed as the homestead. In other states, homestead is a question of fact, and a formal legal claim is not required. Whatever is necessary to establish homestead, clear title to oil and gas interests may not be effectively created in homesteaded property without joinder of both spouses, even where the record shows title in one spouse.

3. *Community Property* —In eight states, including Texas, Louisiana, California and New Mexico, community property statutes create a kind of marital partnership in property acquired during marriage. In community property states, each spouse is presumed to own one-half of all property acquired by either spouse during marriage. Though exceptions are made for property acquired by one spouse by inheri-

tance or with assets owned before marriage, the presumption is strong that property acquired during marriage is subject to the other spouse's right. As with dower and homestead rights, community property rights may not be noted on the record; i.e., the property may appear to be wholly owned by one spouse.

The frequency of marital rights has led to a common practice of execution of documents creating oil and gas interests by both spouses, even where the record shows ownership only by one spouse. A better practice is for the spouse who claims no interest, or only a dower or homestead interest, to sign the instrument specially; i.e., the lease or deed should show only the record owner as grantor and the spouse should sign solely to release any rights of dower, homestead, or other marital interest in the premises. Otherwise, arguments may be raised that the joining spouse is entitled to a share of payments provided for under the document; e.g., in the event of a divorce.

C. DEBTORS/CREDITORS

Though in many states secured creditors (e.g., mortgagees or deed of trust beneficiaries) are considered to hold legal title to the property subject to their claims, there is no doubt but that the right to create and transfer oil and gas rights belongs to the debtor. Whatever the legal fiction as to the state of title, the security interest is limited to protection of the creditor's right to collect his money.

However, both grantors and grantees of oil and gas rights need to look closely at the terms of security instruments. The terms of the mortgage, deed of trust or other document creating the security interest may make the acquiescence of the creditor essential. Many mortgages and deeds of trust contain "due on sale clauses;" e.g., "if the ownership of any portion of the premises shall be changed . . . then, at the mortgagee's discretion, the entire indebtedness secured hereby shall become immediately due and payable". Others contain assignments of proceeds; e.g., "there are specifically assigned to the mortgagee all rents, revenues, damages and payments . . . on account of any and all oil, gas, mining and mineral leases, rights or privileges of any kind now existing or that any may hereafter come into existence." Such provisions may not be enforceable in some states, but they are certain sources of dispute between oil and gas interest owners and holders of security interests. Moreover, unless the secured party consents to the grant of the lease or other interest, foreclosure of a prior secured interest will extinguish the oil and gas interest conveyed. As a matter of practice, grantors and grantees of oil and gas interests commonly seek *waivers of priority* or *subordination agreements* from secured interest owners.

The position of the vendor under a land contract or contract for deed is different only as to procedure. The traditional remedy of the land contract vendor upon default has been to repossess the land and to declare the vendee's equitable title to have been extinguished. Since it is the vendor who will reacquire the full title in the event of default, a ratification of the lease with a present grant of after-acquired rights is usually sought instead of a waiver or a subordination of the vendor's rights.

D. FIDUCIARIES/BENEFICIARIES

Another special situation of joint ownership occurs when a fiduciary holds title or exercises rights of management for a beneficiary. Common law fiduciaries did not have the right to lease for oil and gas development or to create other oil and gas interests because of the traditional view of waste that proscribed any change in the state of the property. Many states have changed that rule for oil and gas leasing, and most trust instruments specifically confer upon the trustee the right to lease and create other oil and gas rights. Therefore, as a general rule, fiduciaries are able to create and transfer oil and gas rights, even though these rights may extend beyond the term of the fiduciary relationship.

The powers of a fiduciary are often subject to statutory conditions and limitations that vary from state to state. For example, although a guardian of a minor has a statutory power to grant a lease on behalf of his ward in Texas, the same statute limits the primary term of any lease to five years. Where the fiduciary relationship is created by trust instrument, similar variations are possible because the terms of the trust prevail over more liberal statutory provisions.

As a result, leases and other interests granted by fiduciaries are frequently the source of title problems. Meticulous attention to the terms of the instrument establishing the fiduciary relationship and to the requirements of state law is essential.

E. EXECUTIVE/NON–EXECUTIVE OWNERS

The executive right is the power to lease minerals. Frequently, the executive right is severed from the other incidents of mineral ownership. For example, O might convey to A, reserving to himself half the minerals and the exclusive right to lease all of the minerals. By so doing O could maintain better control over development. O would have half the mineral interest and the executive right to A's half non-executive mineral interest. To obtain a valid lease, a prospective lessee would have to deal with O, not A.

The executive right is just one of the incidents of mineral ownership. Except in Louisiana, it cannot be held independently of ownership of some oil and gas interest. Generally, it does not entitle its holder to the portion of lease benefits accruing to non-executive mineral interests. In Louisiana, the executive right owner is entitled to bonus and delay rentals, but in other states non-executive mineral interest owners retain the right to all lease payments accruing to their interests.

[95]

It is unclear whether the executive right includes the power to conduct operations on the land as well as lease it. There is a division whether the executive has the power to pool the non-executive rights. In Texas, it has been held that he has not, but Louisiana has permitted pooling.

A frequent source of dispute is the duty owed by the executive to the non-executive. What obligation does O have to A in the example above to exercise the power to lease? Can he decline to lease on any terms? Article 109 of the Louisiana Mineral Code says he can, but *Federal Land Bank of Houston v. United States*, 168 F.Supp. 788 (Ct.Cl.1958), held squarely to the contrary. Another issue is what obligation O owes A to negotiate a "good" lease. Some cases have indicated there is no duty, because O's self interest will protect A. However, most recognize some duty, though it is expressed in a variety of terms. Probably the standard is one of the objective, reasonable prudent mineral owner.

F. LIFE TENANTS/REMAINDERMEN

The most common successive interests, where ownership is divided between present and future rights, are those of life tenants and remaindermen. Typical problems in dealing with life tenancies and remainder interests are: (1) the power to grant; (2) division of proceeds; and (3) the open mine doctrine.

1. POWER TO GRANT

a. In Common Law States

At common law, neither a life tenant nor a remainderman can develop oil and gas, grant a valid oil and gas lease, or create any other oil and gas interest without permission of the other because neither possesses the full rights to the property. The life tenant has the right of present use, but must conserve the estate for the remainderman. He cannot grant an oil and gas lease, for example, because taking minerals would diminish the estate that he must conserve and because the term of the lease (typically "so long as oil and gas are produced") might exceed the life estate. On the other hand, the remainderman will eventually have full rights to the property, but he lacks the right to present use that any interest in oil and gas will require.

If the life tenancy is created by an instrument, rather than by operation of law, the relationship of the life tenant and remainderman may be changed either specifically or by inference. For example, if the life tenant is specifically given the right to lease or otherwise dispose of the property, the weight of authority is that the life tenant has the right to grant an oil and gas lease even though it may extend beyond his lifetime. The right to lease may be inferred from a grant of a life tenancy in the minerals; if the intent was to give the life tenant use of the minerals, that intent requires the right to develop or lease for development.   On the other hand, if the instrument merely creates a life estate "without impeachment for waste", the life tenant has no duty to conserve the minerals against depletion under an oil and gas lease, but there is an unanswered question as to his right to grant lease rights beyond his lifetime.

b.   In Louisiana

In Louisiana, the analogous interests to life tenant and remaindermen are the usufruct and naked owner.   However, the usufruct has no right to take minerals, to lease, or to share the benefits of leasing as a general rule.   The usufruct is entitled only to the benefit of the use of the surface unless the instrument creating it provides otherwise.   The naked owner has all rights to oil and gas.

c.   Common Leasing Practice

Generally, grants of oil and gas rights from life tenants and remaindermen are obtained over the signatures of both.   An oil and gas lease may be obtained (1) by having the life tenant and remainderman sign the same lease, (2) by having the life tenant and remainderman sign separate leases, or (3) by having the life tenant grant a lease which is then ratified by the remainderman.   The third practice is preferred by oil companies for two reasons.   First, it avoids questions as to how payments under the lease are to be divided; the life tenant is designated as the lessor to whom payments are to be made and the remainderman ratifies the lease terms. Second, a remainderman presented with a ratification is less likely to demand payment of bonus or a share of lease proceeds than if he is asked to execute a lease.

Each of the three approaches may lead to problems. If the life tenant and the remainderman execute the same lease, there will be an ambiguity as to how the bonus, delay rentals and royalties provided for in the lease are to be paid, unless the division is spelled out. As is discussed at page 194, ambiguities over how delay rentals are to be divided may result in termination of the lease. Where the life tenant and the remainderman execute separate leases, the lessee may have to make double lease payments unless the leases are carefully drafted. Some courts have held that the lessee must pay whatever bonus, delay rentals and royalty each lease provides for the lessor it names. The third alternative, obtaining a lease from the life tenant and a ratification from the remainderman, may lead to the transaction being set aside for fraud, misrepresentation, or overreaching if its nature is not disclosed. Neither the life tenant nor the remainderman have the right to develop oil and gas without the other. Therefore, the ratification form presented to the remainderman will contain words presently granting the remainderman's future interest to the lessee under the term of the lease being ratified, as well as ratifying the life tenant's lease.

As such, it is more than a mere ratification, and that should be disclosed.

2.   DIVISION OF PROCEEDS

Where a life tenant and remainderman grant an oil and gas lease without agreeing specifically upon division of proceeds under the lease, how should the proceeds of the lease—the bonus, delay rentals, royalty and shut-in royalties—be paid?

The courts have generally allocated funds between life tenant and remainderman on the basis of their classification as income or corpus. If classified as income, money is paid to the life tenant. If classified as corpus, a return of the "body" of the trust, it is invested to yield income (which is paid to the life tenant) and held as principal to be turned over to the remainderman upon the life tenant's death.

When applied to oil and gas lease proceeds, application of that rule rarely pleases either life tenant or remainderman. Delay rentals (which have traditionally been a nominal dollar per acre per year) are uniformly classified as income and paid to the life tenant. Bonus payments, to induce grant of the lease, and royalty payments are usually allocated to principal and invested. The interest from investments is paid to the life tenant, but the remainderman gets nothing until the life tenant dies. In Arkansas and Oklahoma, bonus has been allocated to the life tenant. But generally, the life tenant is paid only delay rentals and the interest on bonus and royalties, and the remainderman receives nothing until the life tenant's death. Often, the life tenant and remainderman will agree in advance upon allocation of lease proceeds; if they agree, all of the proceeds can be distributed.

3. THE OPEN MINE DOCTRINE

The open mine doctrine, borrowed from the law of hard minerals, changes the general rules for division of oil and gas lease proceeds. Where there is an "open mine" on the property when the life tenancy is created, the life tenant is entitled to all payments under a lease, including bonus and royalties (as well as the right to work the mine in absence of a lease). The rationale is the presumed intent of the life tenancy's creator that the life tenant should have the use of the property as it was when the life tenancy was created.

Generally, a mine is held to be "open" where there is an oil and gas lease in existence at creation of the life tenancy; the grant of a lease opens the mine. Contrary to the rule for hard minerals, it appears that where there is production from a well at creation of the life tenancy, the life tenant is also entitled to the proceeds from additional wells drilled. In Texas and Oklahoma at least, the open mine doctrine is limited to the term of the lease in existence when the life tenancy is created. The life tenant may not grant additional oil and gas leases on the property or extend existing leases.

The Louisiana Mineral Code adopts a version of the open mine doctrine as an exception to the general rule that the naked owner is entitled to the benefits of leasing. If when the usufruct is created there is a well capable of production on the land or on land pooled with it, then the usufruct is entitled to royalties on actual or constructive production. Further, the usufruct has the right to lease the interest subject to the usufruct and to retain bonus and rentals.

G.   TERM INTERESTS

The theoretical position of a holder of an estate for years is the same as a life tenant. He lacks the power to grant an oil and gas lease because development of petroleum would be waste and because the lease may be extended beyond his lifetime by production. He should be benefitted by the open mine doctrine. In practice, grants of estates for years are rarely intended to include the minerals as well as the surface.

However, defeasible term interests in oil and gas are frequently seen. For example, O may convey to A mineral rights "for 10 years and so long thereafter as oil or gas are produced . . . ." Such language gives A rights that will terminate at the end of the 10 years without production, but that will be extended by production as long as production lasts.

Defeasible term interests present the same interpretative difficulties as the term clause of an oil and gas lease. Should principles and precedents applied to leasing govern defeasible term interests? In Kansas and Texas, courts have held that they should because the likelihood is that the parties intended that result by their choice of language. Oklahoma has rejected the notion, adopting Professor Eugene Kuntz' analysis that leases contemplate development by the lessee while defeasible term interests are held for speculation. Accordingly, in *McEvoy v. First National Bank & Trust of Enid*, 624 P.2d 559 (Okla.Ct.App.1980), a defeasible term interest for twenty years and so much longer as oil or gas were "produced in paying or commercial quantities" terminated where there was a well on the premises capable of production but not actually producing at the end of the term. Capability of production would have preserved a lease with similar language. Louisiana's Mineral Code also applies a different standard to interruption of prescription for mineral servitudes than for oil and gas leases. However in contrast to Oklahoma, it sets a lower standard for servitudes than for leases. Article 38 makes good faith operations sufficient to interrupt prescription of mineral servitudes, while actual production is required for leases.

Similar problems are presented with termination of defeasible term interests. Does a reference to "production" require "production in paying quantities," as it generally does in oil and gas leases? Should whether a cessation of production is temporary or permanent be judged by the same factors applied to leases? Again, authority is divided. However, the reasons for the development of the rules for leases do not apply to defeasible term interests. Lease production must be "in paying quantities" because the lease is given and taken with an expectation of profit from development, and the lessee has it within his control to make operations profitable. Defeasible term interests are often held for speculation and their owners usually lack the right and the expertise to develop themselves. For these reasons, the standard by which a temporary cessation of production is judged should be more lenient for defeasible term interests than for lease interests. However, except in Kansas, Oklahoma, and Texas, there is little case law on these issues.

CHAPTER 7

INTERPRETIVE PROBLEMS IN OIL AND GAS CONVEYANCING

Conveyances of oil and gas interests frequently present problems of interpretation. Those problems are often dealt with by well-established rules of judicial construction that yield little certainty or, where they are certain, may seem unfair in result. In this chapter, we will examine common conveyancing problems with the purpose of identifying them and learning to avoid them.

A. STEPS IN JUDICIAL INTERPRETATION

The first duty of a court confronted with interpretation of a conveyance is to give effect to the intention of the parties. Though certainty of titles is desirable, preservation of ownership is the controlling policy. To ascertain the intention of the parties, the courts have developed a three step process for interpreting conveyances of real property interests, including oil and gas rights:

(1) determine the intention of the parties from the terms of the instrument;

(2) if the intention of the parties is doubtful, use construction aids or rules of construction to ascertain their objective intent;

[105]

(3) if the instrument is ambiguous, consider parol or other extrinsic evidence.

1. INTERPRETATION OF THE INSTRUMENT AS A WHOLE

A court's first step in interpreting an instrument is to look to all of its terms. Courts seek the intention of the parties, but because of the importance placed on certainty in real property ownership, intention is sought in an objective manner by examining the terms of the instrument rather than by asking the parties what they intended. The instrument in question is reviewed as a whole, and an attempt is made to reconcile all of its terms. This first step is frequently referred to as the "four corners rule", because the court looks to the four corners of the instrument to ascertain the parties' intention.

2. USE OF CONSTRUCTION AIDS OR CIRCUMSTANTIAL TESTS

Where there is doubt about the intent of the parties to the instrument after examination of its four corners, the courts apply a variety of construction aids and circumstantial tests. For example, the instrument will be construed against the interests of the party who prepared it, who was in a position to have made its intent clear. Typed or handwritten provisions will prevail over printed provisions, because they are more likely to express the intention of the parties. General terms following specific terms will be interpreted by the rule of *ejusdem generis* to refer to terms of the same kind or class as the specific terms.

Construction aids and circumstantial tests provide no certainty that the intention of the parties will be given effect. At best, they provide an objective inference of what their intention might reasonably have been. Many, like the rule that an instrument is construed against the party who prepared it, further policy goals unrelated to intent.

3.   CONSIDERATION OF EXTRINSIC EVIDENCE

It is frequently impossible to find a clear inference of intent of the parties from the terms of the instrument even with the help of construction aids. For example, as is discussed at page 109, a mineral deed form conveying "oil, gas and other minerals" gives little guidance as to what other substances the parties may have intended to include within the phrase "other minerals." Therefore, as a last resort, the courts examine the attendant circumstances of the conveyance. Parol (oral) evidence may be considered, as may extrinsic evidence in the form of letters, memoranda, or records bearing upon the negotiations that led to the ambiguous conveyance, and the performance by the parties before the dispute arose.

4. APPLICATION OF THE INTERPRETIVE STEPS

Knowing the steps courts follow in interpreting conveyances is useful in understanding the process of judicial decision making. However, it is of little help in predicting the result of particular disputes. Since the intention of the parties to the conveyance is the interpretative goal, determination of what the instrument fairly says, what construction aids to use, and what extrinsic evidence is sufficient to establish intent is within the discretion of the courts. As a result, litigation is the only certain manner of resolving many interpretative problems of oil and gas conveyancing.

B. WHAT IS THE MEANING OF "MINERALS"

The problem of what substances are included in a grant or reservation of "minerals" is a good example of the difficulties of applying the principles of judicial interpretation. The meaning of a general reference to "minerals" in a grant or reservation is one of the most economically significant issues of the century for the natural resources industry. Commercial uses have been developed for many substances that had little or no commercial value a generation ago. In addition, reserves of many important natural resources are growing short, so that what used to be marginal deposits have assumed substantial value. As a result, there is frequent litigation over ownership of substances that may have been granted or reserved as "minerals."

The problem does not often arise with fugacious substances. Except in Pennsylvania, the courts have held that oil and gas are "minerals." Helium produced in conjunction with natural gas has been held to be "gas," even though it is not combustible. However, the dispute is common with "hard" minerals. Uranium is a good example. Suppose that the record of title to Blackacre shows that O, as owner in fee simple absolute, conveyed the land in 1950 to A, reserving the "oil, gas and other minerals." There have been subsequent transfers of both the surface interest and the severed mineral interest so that ownership is now vested in P, who owns the "oil, gas and other minerals" and D, who owns the remainder of the interest in the land. If a mining company wishes to obtain the right to strip mine uranium from the premises, should it take a lease from P or D? Is uranium a "mineral" or does it belong to the surface owner?

1. WHAT THE COURTS HAVE DONE

When confronted with such issues, the courts have generally said that a reference to "minerals" or "other minerals" following reference to oil and gas is uncertain. Attempts to ascertain the intention of the parties from the four corners doctrine have proved unsuccessful. Ownership of the substances has been established by application of a variety of circumstantial tests and construction aids. In addition to construction against the grantor, the following have been applied:

ejusdem generis —general words that follow specific words are limited to things of the same kind or classes as those specifically stated, so that a reservation of "oil, gas and other minerals" reserves other minerals "like" oil and gas;

community knowledge test —the substance is considered to be a "mineral" if it was regarded as such by the community in which the instrument was given at the time of the conveyance;

exceptional characteristics test —a substance is a "mineral" if it possessed exceptional characteristics that gave it special value at the time of the conveyance;

rule of practical construction —the actions of the parties contemporaneous with and subsequent to the conveyance are considered to establish the intention of the parties. If the negotiations concerned only oil and gas rights, only substances produced in conjunction with oil and gas are likely to be "minerals;"

surface destruction test —where production of a substance requires destruction of the surface, the substance is not a "mineral" because the original parties would not have intended that the mineral interest owner be given the right to destroy the beneficial use of the property by the surface owner.

Application of the variety of circumstantial tests and construction aids yields diverse results. For example, a grant of "coal and other minerals" was held to include sand and gravel in a case in which the court applied the rule of practical construction but not to include sand and gravel in another case where the exceptional characteristics test was applied. Furthermore, courts frequently apply several construction aids or circumstantial tests in a single case. The result is confusion and uncertainty.

2. THE TEXAS EXPERIENCE

a. *Acker* v. *Guinn*: The Surface Destruction Test Articulated

The Texas experience with the meaning of "minerals" is a good example of the difficulties courts have had with the issue. In *Acker v. Guinn*, 464 S.W.2d 348 (Tex.1971), the Texas Supreme Court held that low grade iron ore was not a "mineral" in a conveyance of an undivided interest "in and to all the oil, gas, and other minerals." The result was reached by application of a surface destruction test. The court reasoned that the grantee and grantor to a severance of minerals do not contemplate that the utility of the surface for agricultural or grazing purposes will be destroyed. Therefore, unless the intention of the parties is clearly otherwise, grant or reservation of "minerals" does not include substances that must be removed by surface destructive techniques.

[*111*]

The decision in *Acker v. Guinn* caused substantial confusion for two reasons. First, it did not make clear what date was to be used to apply the tests; e.g., the date of the instrument, the date the matter was referred to court, or the date mining was to be undertaken. Second, it did not make clear how surface deposits that extended to depths that would normally be developed by shaft mining would be treated; whether they belonged solely to the surface owner, or to the mineral owner, or their ownership would be apportioned.

b. *Reed* v. *Wylie* I: The Surface Destruction Test Redefined

The Texas Supreme Court addressed these questions six years later in *Reed v. Wylie*, 554 S.W.2d 169 (Tex.1977). The grantors had reserved a fractional interest in "minerals" and sought to strip mine lignite. In holding that lignite belonged to the surface owner, the court said that what substances were minerals was to be determined as of the date of the instrument. If at that date a substance had been located so near the surface that extraction would necessarily have consumed or depleted the land surface, then deeper deposits belonged to the surface owner also.

Reed v. Wylie answered the questions raised by *Acker v. Guinn*, but it spawned its own. It held that the surface owner and not the mineral owner retained the title to substances "at the surface." But how close to the surface was "at the surface?" Must the minerals be at the surface on the land in question, or would it be sufficient if they were at the surface nearby, though not on the land? Another series of problems arose because of the necessity of establishing whether the method that would have been used to extract particular kinds of minerals would destroy the surface. *Reed v. Wylie* required that the surface owner show that the only feasible methods of recovering the disputed substance at the time of the grant or reservation were surface destructive. This required courts to ascertain conditions at the time of the deed, and meant that the term "minerals" might have different meanings in identical instruments given at different times. Further, there were questions about application of the surface destruction test. Did substances belong to the surface owner if the *only* way to get them was surface destructive or if any one of several methods was surface destructive? Should courts consider only production techniques in actual use, or should pilot or experimental techniques be considered also? And how broad an area should be surveyed to determine methods available to remove the substance in question—the locality, the State of Texas, the nation?

c. *Reed* v. *Wylie* II: Further Refinement

In 1980, the issue of what substances are "minerals" again came before the Texas Supreme Court. In *Reed v. Wylie,* 597 S.W.2d 743 (Tex.

1980), the court again modified its position. It said that the surface owner need not show that the *only* methods of extraction available were surface destructive; it would be sufficient to show that "any reasonable method" was so. It said that the surface destruction test was one of current impact. What production techniques were available should be determined as of the time the court makes its decision rather than when the deed was executed, so as to avoid difficulty of proof. Finally, it said that its reference in *Reed* I to "at the surface" meant merely "near the surface." Applying these modifications, the court again ruled that the lignite belonged to the surface owner.

The 1980 *Reed v. Wylie* decision apparently left the following formulation of the law in Texas:

A substance is not a "mineral" within the meaning of a general grant or reservation of "minerals", if substantial quantities of that substance lie at or near the surface in the reasonably immediate vicinity so that one of the reasonable methods of its removal when the matter is decided by the court would be by strip or open pit mining. If a substance is not a "mineral", the surface owner retains ownership of it at whatever depth below the surface it may be found.

But the 1980 decision raises questions of its own. When are deposits "in the reasonably immediate vicinity?" When are deposits "at or near the surface?" Must the deposits conform generally to the contour of the earth's surface? What is a reasonable method of recovery? How substantial must the deposits be; e.g., what if there is a small deposit of uranium at the surface and a much larger ore deposit deeper? Under the current impact test, will ownership of substances shift between the parties as new technologies are developed or new deposits found? These issues will require more litigation.

3.   A PROPOSED SOLUTION

The Texas cases illustrate that trying to formulate a rule that allocates ownership of substances between the surface owner and the mineral owner on the basis of rules of construction or interpretive aids inevitably opens the proverbial can of worms. A better approach would be to separate the issues of ownership and enjoyment. Professor Eugene Kuntz has said it best:

> "[T]he courts are seeking to give effect to an *intention* to include or exclude a *specific substance*, when, as a matter of fact, the parties had nothing specific in mind on the matter at all. . . . The intention sought should be the *general intent* rather than any supposed but unexpressed specific intent, and, further, that general intent should be arrived at, not by defining and redefining the terms used, but by considering the *purposes* of the grant or reservation in terms of the manner of enjoyment intended in the ensuing interests."

"When a general grant or reservation is made of all minerals without qualifying language, it should be reasonably assumed that the parties intended to sever the entire mineral estate from the surface estate, leaving the owner of each with definite incidents of ownership enjoyable in distinctly different manners. The manner of enjoyment of the mineral estate is through extraction of valuable substances, and the enjoyment of the surface is through retention of such substances as are necessary for the use of the surface. . . . "

"Applying this intention, the severance should be construed to sever from the surface all substances presently valuable in themselves, apart from the soil, whether their presence is known or not, and all substances which become valuable through development of the arts and sciences." Kuntz, *The Law Relating to Oil and Gas in Wyoming*, 3 Wyo.L.Rev. 107, 112–13 (1947), reprinted 34 Okla.L.Rev. 28, 33–34 (1981) (Emphasis in original).

Professor Kuntz' analysis has been termed the "manner of enjoyment" theory. It has been cited favorably by various courts, but rarely applied to solve the interpretive problem of a general grant or reservation of "minerals."

Probably the major reason that a broad definition of minerals has not been embraced by the courts is the assumption that its adoption would give the mineral owner the right to destroy the surface (albeit with compensation) if that were necessary to get the minerals. That result is not logically required. The mineral owner's ownership of all substances that have or may acquire value apart from the soil can be recognized while his right to use of the surface is limited to techniques that are not surface destructive. Under such an approach, the mineral owner would be forced to negotiate an accommodation with the surface owner or to await (or develop) methods of extraction that would not destroy the surface.

This approach would not abolish litigation and uncertainty relating to grants or reservations of "minerals" or "other minerals." However, it would shift the focus from the illusory specific intention of the parties to the reasonableness of operations on the surface, and that would be a substantial step forward from where we are now.

C. THE MINERAL/ROYALTY DISTINCTION

1. THE SIGNIFICANCE OF THE DISTINCTION

From the discussion in Chapter 3, it should be apparent that whether an interest is a mineral interest or a royalty interest is an important distinction. A mineral interest possesses the right to develop or to lease, and to keep the proceeds of leasing. A royalty interest lacks those rights, but has a right to a share of production free of the costs of production. Which is to be preferred depends upon the circumstances. Even if there is never production from the land, the mineral interest may have substantial value because it has the right to lease and to keep bonus and delay rentals. On the other hand, if there is production, a royalty interest is usually preferable to an equal mineral interest because it is cost free.

2.   COMMON INTERPRETIVE PROBLEMS

Interpretive disputes about whether an interest is a mineral interest or a royalty may arise either from the language of the conveyance or from the general situation. For example, to illustrate both, suppose that at a time when O owned the fee simple absolute and there were no oil and gas leases outstanding, the record showed a conveyance from O to A of "$^1/_2$ of the royalty in the oil, gas, or other minerals in and under the said above-described lands." Must an oil and gas lease be granted by A in order for a lessee to acquire full operating rights?

There is no clear cut answer. At first glance, the language in the example may look like it creates a royalty interest, because it refers specifically to "royalty." On the other hand, the remainder of the description "in the oil, gas or other minerals in and under," is more consistent with an attempt to describe the minerals in place than an interest in production if and when it occurs. Furthermore, a 50% royalty interest, one that would give its holder the right to 50% of all production cost free, is an unusual burden. An inference may arise from the size of the fraction

that the parties intended to give a mineral interest. Another possibility is that the reference to
50% *of* the royalty is to 50% of any royalty that
may be provided for in a future oil and gas lease.
However, there was no lease on the property
when the conveyance was made and the language
does not refer to future leases.

The major reason that the mineral/royalty distinction is so hard to deal with is that the courts
turn themselves inside out trying to ascertain the
intent of the parties. The result, as with the
problem of what is a "mineral", is apparent inconsistency in the decisions.

a.   Guidelines to Interpretation

Because of the diversity of precedent from
state to state and even within certain states, it is
difficult to generalize as to when ambiguous language will be effective to create a mineral interest or a royalty interest.

No one factor in the instrument is determinative to establish the intention of the parties to
create a particular interest. But a specific statement of intention in the granting clause or in a
clause immediately following will probably be
given more weight than any other factor, including the words of grant. *Atlantic Refining v.
Beach*, 78 N.M. 634, 436 P.2d 107 (1968), is a good
illustration. There, a document headed "Mineral
Deed" mixed mineral interest and royalty references. In the granting clause it conveyed an un-

divided $^1/_{16}$ mineral interest, but in a clause following there was stated an intention to retain all rights to delay rentals and to convey "one half of the royalty." The Supreme Court of New Mexico affirmed the trial court and held that the deed reserved a royalty interest of $^1/_2$ of the usual $^1/_8$, giving effect to the intention language of the deed rather than to the literal terms of the granting clause. The court considered the title of the instrument, the language of the grant, and the reference to easements of access and egress for development, as well as the statement of intention. No one factor was determinative but the intention statement was of prime importance.

Use of the term "royalty" is generally considered not to indicate clearly an intention to create a royalty interest unless there is other language in the conveyance that suggests an intention to use the word in a technical sense. In the vernacular, "royalty" is often used synonymously with "mineral interest." In Oklahoma, where there is no lease in existence at the time of the conveyance and the royalty is not stated as a specific percentage of production (e.g., "$^1/_2$ of $^1/_8$ royalty"), cases hold that reference to royalty denotes a mineral interest. Reference to royalty also created a mineral interest in *Corlett v. Cox*, 138 Colo. 325, 333 P.2d 619 (1958). The Colorado Supreme Court relied on a now defunct common law rule that a grant of the rents or profits of land is a grant of the land itself. It held that a reservation of "$6^1/_4$% of all gas, oil and minerals that may be produced on any or all of the above mentioned land, or in other words . . . $^{1}/_{2}$ of the usual $^{1}/_{8}$ royalty" created a mineral interest. As a result, it is questionable whether a royalty interest independent of a lease can be created in Colorado. In contrast, use of terms such as "minerals," "mineral interest," or "oil and gas rights" are likely to be held to create mineral interests unless they are inconsistent with other provisions of the instrument.

Many instruments attempt to define the nature of the interest rather than to give it the title of "mineral interest" or "royalty interest". This is usually an effective method of description, unless conflicting descriptions are mixed. References that describe oil and gas in the ground (e.g., "$^{1}/_{32}$ of the oil and gas in and under" or "in and under and that may be produced from") are likely to be interpreted as mineral interests, even in states that follow the non-ownership theory where they cannot be given literal effect. References that seem to convey an interest in oil and gas after it is produced (e.g., "$^{1}/_{32}$ of all the oil and gas that may be produced and saved") are likely to be treated as creating royalty interests. However, in several states, that is true only if the language also indicates that the right to production is to be cost free. At least one prominent commentator, Professor Richard Maxwell, suggests that the real distinction between a mineral interest and a royalty is that a royalty is free of costs of production while a mineral interest is subject to expenses. A reference to sharing the costs indicates a mineral interest. Statement that the interest is cost free suggests a royalty.

b. Avoiding Ambiguity

Use of commercially printed form deeds will not necessarily avoid the problem of the mineral/royalty distinction. Most royalty deed forms (but not the one included in the Appendix) contain language granting easements for "the right of ingress and egress at all times *for the purpose of mining, drilling and exploring* said lands for oil and gas and other minerals and removing the same." Many title attorneys believe that such language is more appropriate for a mineral deed (because a royalty interest has no right to operate) and declare such a royalty deed form ambiguous. Most mineral deed forms describe the interest to the oil and gas and other minerals "in and under *and that may be produced from.*" Though such language is accepted as creating a mineral interest, its use has occasionally led to dispute on the grounds that it mixes references appropriate to mineral and royalty interests.

Draftsmanship by reference to the incidents of the interest created rather than by use of terminology is the best assurance of avoiding ambiguity. For example, if a royalty interest is sought, reference to the cost-free nature of the interest and to its lack of operating or leasing rights is desirable. "Mineral interest" and "royalty interest" are shorthand terms devised by the courts to describe the most common bundles of rights that parties to a severance wish to create. Other groupings are possible. By referring specifically to the individual rights that make up the interest, the instrument can make clear whether there is intended a mineral interest, a royalty interest, or some hybrid of the two. The royalty deed form in the Appendix is a good attempt at drafting by reference to characteristics.

D. FRACTIONAL INTEREST PROBLEMS

Fractional interests are more commonly found with oil and gas interests than with any other property. As a result, lawyers and landmen, many of whom went to law school or into land management to avoid the mathematics required for the sciences, find themselves working with fractional interests that typically extend to seven decimal places.

Often they do not handle the creation of such interests well. Disputes as to how much is conveyed are common. In this section, we will review three common problems that arise with conveyancing of fractional interests.

1. DOUBLE FRACTIONS

Frequently, when one who owns less than all of the mineral interest or royalty interest conveys or reserves a fraction, there is ambiguity as to whether the grant or reservation is intended to be a fraction of the whole or a part of the fraction

owned by the grantor. This is called the double
fraction problem.

Suppose, for example, that:

O owns the surface and an undivided $\frac{1}{4}$ mineral interest in Blackacre; and

O conveys to A "all of my right, title and interest in Blackacre, reserving an undivided $\frac{1}{4}$
of the minerals in and under said land";

What does O retain?

There is an inherent ambiguity. Ordinarily, a
grant or a reservation of a fractional interest will
be considered to be a grant or reservation of the
whole. However, where there is an expressed or
implied reference to a fraction, as there is in the
example because of the reservation from "said
land," which is a reference to "all of my right,
title and interest," there is an ambiguity. Is the
reservation of $\frac{1}{4}$ of all of the minerals or of $\frac{1}{4}$ of
the $\frac{1}{4}$ that the grantor had? The same problem
would have been created had O described the
property being transferred by reference to the
prior deed which conveyed only a fractional interest.

a. Guidelines to Interpretation

The courts seek to solve the double fraction
problem by the steps of judicial interpretation
previously considered at pages 105–107. The results reached are no more pleasing than those
reached with respect to the issue of what substances are "minerals" or the mineral/royalty dis-

tinction. Often, the result is reached by a narrow construction of the deed that seems neither certain nor fair. For example, in *Spell v. Hanes*, 139 S.W.2d 229 (Tex.Civ.App.1940) (error dismissed, judgment correct), the grantor, who owned an undivided ⅝ interest in the minerals, conveyed "an undivided ¼ interest in and to all of the . . . minerals", and described the land. Following the grant, the phrase was added "Above grant is to apply to our undivided interest in and to the above described lands." That language suggests to this writer an intention to grant ¼ of his ⅝ths interest. Despite the intention language, the court held that the deed conveyed ¼ of all the minerals. The court gave effect to the granting clause, saying that the intention language merely indicated what interest the grant should be charged against.

Black v. Shell Oil Co., 397 S.W.2d 877 (Tex. Civ.App.Ct.1965), is another example. The grantors, who owned an undivided ½ mineral interest, conveyed "an undivided one-half (½) interest" in minerals, "It being the intention of grantors herein to convey one half of the minerals *out of* the interest owned by them." (Emphasis added). The grantor's successors contended that the deed conveyed ½ of the ½ the grantors had. However, the court held that the deed was not ambiguous, that "out of" designated the interest from which the ½ mineral interest was to be taken. It noted that reference to "of" instead of "out of" would have achieved the result urged. Such results are correct as a literal interpretation of the words used, but they leave one with an uncomfortable feeling that the grantor's intention was not achieved.

b. Avoiding Ambiguity

Good draftsmanship will avoid the double fraction problem. Whenever one who owns a fraction conveys or reserves a fraction, the grant or reservation should specify how it is to be measured. In our example, there would have been no ambiguity had O reserved "$1/4$ of 100% of the minerals" or "$1/4$ of $8/8$ of the mineral interest," or had O conveyed to A the surface only. Likewise, there would have been no ambiguity if O had reserved "$1/4$ of the portion of the mineral interest that I own."

2. OVERCONVEYANCE

A second frequently encountered problem in conveyances of fractional oil and gas interests is overconveyance; transactions in which the total of the fractions reserved and conveyed is greater than 100%. It may be helpful to think of this ambiguity as *the sum of the parts is greater than the whole* problem.

An illustration may help. Suppose that:

X conveys Blackacre to O, reserving an undivided $1/4$ mineral interest; and

O conveys Blackacre to A by warranty deed, "reserving to O an undivided ¼ of the mineral interest";

What does O retain?

The issue is whether A takes the surface and ½ of the mineral interest, or the surface and ¾ of the mineral interest, or (perhaps) the surface and ⁹⁄₁₆ of the mineral interest. Clearly A takes subject to X's retained ¼ mineral interest if X's deed to O was recorded. Clearly, as well, O's *intent* was to retain ¼ mineral interest for himself; his reservation says so. But the effect of the grant of the described property by warranty deed is to guarantee that 100% of the property described is transferred, except for any interest specifically reserved. If that rule is applied here and effect given to O's intent as well, there will be an overconveyance (A's ¾ + X's ¼ + O's ¼ = 125%); the sum of the parts will be greater than the whole.

a.   The *Duhig* Rule

Courts generally deal with the problem of overconveyance by warranty deed by deducting the overconveyance from the grantor's interest, to the extent that is possible, by application of what is called the *Duhig* rule, after the case of *Duhig v. Peavy-Moore Lumber Co. Inc.*, 135 Tex. 503, 144 S.W.2d 878 (1940). That rule may be summarized as follows:

Where full effect cannot be given both to the granted interest and to a reserved interest because of a previous outstanding interest, priority will be given to the granted interest (rather than to the reserved interest) until full effect is given to the granted interest.

Thus, in our illustration (based on *Body v. McDonald*, 79 Wyo. 371, 334 P.2d 513 (1959)), O retained nothing despite his clear intent. The rule has been adopted specifically or in effect in Texas, Colorado, Oklahoma, New Mexico, Wyoming, Louisiana, Mississippi and Alabama.

As adopted by the Texas Supreme Court, the *Duhig* rule is an application of an analogy to estoppel by deed. O's conveyance to A is held to be conveyance of the whole by ordinary conveyancing principles. O's reservation of ¼ is deemed to be ¼ of the whole by those same principles. Since giving effect to O's reservation breaches his warranty to A, O is held estopped to assert his reservation to the extent of the overconveyance. Another view of the principle is that the conveyance with covenants of warranty to A shows O's intent to convey the surface and ¾ of the mineral interest, since that is what it literally says. Therefore, O's reservation of ¼ refers to the prior outstanding interest. By either view, the overconveyance is charged against O.

The *Duhig* rule is significant in oil and gas conveyancing because of the element of certainty that it brings to titles. Without such a rule, the ambiguities inherent in our illustration could be solved only by litigation. Where the rule is applied, the record state of title can be relied upon.

b. Departures From the Rule

An unresolved issue is how rigidly the *Duhig* rule should be applied. What if the grantee has actual notice of the unmentioned outstanding interest, for example? Should fairness bar application of the rule? Generally, the rule has been applied as a matter of law. However, in *Gilbertson v. Charlson*, 301 N.W.2d 144 (N.D.1981), it was rejected where the grantee had actual notice of outstanding interests that were not specifically recited in the reservation. The court decided the case on ordinary principles of equitable estoppel, which is based on the actions or statements of the parties rather than the formal representations of the deed. Since the grantee knew or ought to have known of the outstanding interests, she was not misled by the improper warranty. Were similar reasoning applied in our illustration, A would take $\frac{1}{2}$ mineral interest and O's reservation of $\frac{1}{4}$ mineral interest would be given full effect, if A knew of the outstanding interest in X when the conveyance was made.

In another case, *Hartman v. Potter*, 596 P.2d 653 (Utah 1979), the *Duhig* rule was simply ignored. There, the grantor, who owned $\frac{3}{4}$ mineral interest, conveyed by warranty deed and reserved $\frac{1}{2}$ mineral interest. The court found that since the grantee knew of the previously outstanding interest and since the grantor could not grant what he did not own, he reserved $1/2$ of the $3/4$, or $3/8$ mineral interest. Similar reasoning applied to our example would result in awarding A $9/16$ mineral interest and O $3/16$ mineral interest.

Professor Willis Ellis has criticized these decisions as abandoning objective interpretation standards for the dubious equity that consideration of subjective factors may bring. Fair though decisions like *Gilbertson v. Charlson* and *Hartman v. Potter* may be if the grantor has actual knowledge of the previously reserved interest, they undercut the function of warranty deeds in the record title system. The ability of title searchers to rely on the record is key to the operation of the land title system. If title searchers must investigate the knowledge of the grantee, title searches become slower, more expensive, and less certain. In addition, such decisions are inconsistent with the presence of warranties in the deeds. A covenant of title is essentially meaningless if it guarantees only the interest the grantor actually owns.

In states that have recognized the *Duhig* rule, the equitable remedy of deed reformation is available to correct unfairness that results when the rule prevents the intention of the grantor and grantee from being given effect. Even recognizing that reformation will not bring equity in every case (because, for example, it may not be available where the grantee has conveyed to a third party), application of the *Duhig* rule seems preferable to the alternative.

c.   Application to Leases as Well as Deeds

An unresolved issue is whether the *Duhig* rule applies to leases as well as deeds. In *McMahon v. Christmann*, 157 Tex. 403, 303 S.W.2d 341 (1957), the Supreme Court of Texas refused to do so. In a typical leasing transaction, a lessor who owned a $\frac{1}{6}$ mineral interest granted a lease that contained a warranty not limited to his fractional interest and a lesser interest clause that permitted the lessee to reduce lease payments proportionately if the lessor owned less than 100% of the mineral interest. The lease provided for a $\frac{1}{8}$ landowners royalty and an overriding royalty of $\frac{1}{32}$ of oil and gas produced "without reduction." The lessee contended that the lessor was barred by the *Duhig* rule from enforcing the overriding royalty "without reduction," since he had warranted full title but had only $\frac{1}{6}$.

With reasoning that other courts probably will find compelling, the Texas Supreme Court refused to extend the *Duhig* rule to oil and gas leases. It said that oil and gas leases are a special conveyance. The lessor customarily grants a lease of the whole mineral interest though he owns only a fraction, leaving the lessee to reduce payment proportionately by application of the lesser interest clause; i.e. the parties to such transactions do not intend that estoppel apply. Furthermore, the court noted, oil and gas leases are commonly prepared by lessees, not by lessors, so there is no reason to interpret an ambiguity against the lessor.

d.   Avoiding the Overconveyance Problem

Overconveyance (and the *Duhig* rule) can be avoided by drafting.   At least three alternatives are available:

(1) the reservation can specifically refer to all previously reserved or conveyed interests. For example, there would be no overconveyance in our illustration if the conveyance to A provided that it reserved to O ¼ of the mineral interest "in addition to the ¼ mineral interest previously reserved to X" or "excepting all previously reserved interests;"

(2) the reservation may be coupled with a statement of intention that makes it clear what the interests of the parties are to be; e.g. "reserving to O an undivided ¼ mineral interest, it being the intention of the parties that A shall have an undivided ½ mineral interest and that O shall retain an undivided ¼ mineral interest, in addition to the ¼ mineral interest previously reserved to X;"

(3) the grant can be worded to avoid ambiguity; e.g., "O grants A all the surface rights and ½ of all the minerals in and under and that may be produced from" Blackacre.   If the grant is properly worded, there is no need for language reserving an interest.

Of course, more than one of these alternatives may be used in combination.

3.   MINERAL ACRES/ROYALTY ACRES

a.   Mineral Acres

A mineral acre is the full mineral interest under one acre of land. Often grants or reservations are made in terms of "mineral acres" to avoid the ambiguities inherent in fractional interest conveyances. Conveyancing in mineral acres is a useful tool where the intention of the parties is to establish a minimum limitation on the grant or reservation; i.e., the grant of "an undivided 25 mineral acres in Blackacre" is definite and certain.

However, a grant of mineral acres is not necessarily the equivalent of a fractional interest conveyance. For example, if Blackacre is a total of 100 acres, the parties may intend that a grant of "25 mineral acres in and under Blackacre" be the equivalent of an undivided $\frac{1}{4}$ mineral interest. But if Blackacre is either more or less than 100 acres, 25 mineral acres will be, respectively, less or more than an undivided $\frac{1}{4}$ mineral interest.

If references to mineral acres and undivided fractional interests are mixed in a grant or reservation of an interest in property that subsequently turns out to be either larger or smaller than the parties originally anticipated, the stage is set for litigation; e.g. suppose that O, who believes that Blackacre totals 100 acres, conveys "reserving and excepting 25 mineral acres, being an undivided ¼ mineral interest" and a subsequent survey shows that there are 105 acres in Blackacre? Here, reference to mineral acres is in conflict with the reference to the fractional mineral interests; ¼ mineral interest in 105 acres is the equivalent of 26.25 mineral acres. There is an inherent ambiguity that must be resolved.

b.   Royalty Acres

Occasionally, references are seen to "royalty acres." Professors Howard Williams and Charles Meyers define the term as the full lease royalty (whatever percentage may be specified in present or future leases) under one acre of land. Others have suggested that a royalty acre should be seen as the full ⅛ royalty on an acre of land (the "standard" royalty percentage when the relatively few cases on the subject were decided) or as the full production from one acre. Prudence suggests that the term be avoided.

E. CONVEYANCES OF LEASED PROPERTY

Conveyances of property subject to oil and gas leases have frequently led to disputes. Three common problems are (1) the "subject to" problem; (2) apportionment of royalties; and (3) top leasing.

1. THE "SUBJECT TO" PROBLEM

a. Purpose of the "Subject to" Clause

The "subject to" clause in a mineral deed states that the deed is subject to existing oil and gas leases. It has two purposes. First, it protects the grantor against claims for breach of warranty because of the outstanding lease. Second, it is intended to make clear what interest the grantee is to receive in unaccrued rentals and royalties under the lease.

The second goal probably need not be of concern today. It is clear that conveyance of land with an oil and gas lease on the record both binds the grantee to the terms of the lease and entitles the grantee to any unaccrued benefits. However, an early Texas case held that unaccrued lease benefits passed to the grantee only if there was a specific assignment. As a result, the subject to clause was placed in mineral deed forms immediately following the granting clause, rather than as an exception to the warranty, to give the grantee the right to unaccrued benefits. The mineral deed in the Appendix contains a subject to clause in the paragraph following the granting clause.

b. The Two Grants Doctrine

The "subject to" clause is a source of ambiguity when the interests referred to in the clause are inconsistent with those of the granting clause. When that occurs, an ambiguity arises whether the "subject to" provision states an exception to the warranty or describes a second grant, in addition to the one described in the granting clause. A classic case in point is *Hoffman v. Magnolia Petroleum Co.*, 273 S.W. 828 (Tex.Comm.App. 1925), which announced what is called the "two grants" theory. There the lessors, who owned one half the mineral interest in 320 acres subject to an oil and gas lease, conveyed to the plaintiff their mineral interest in 90 of the 320 acres. The granting clause was followed by a subject to clause that referred to the lease and provided specifically that "It is understood and agreed that this sale is subject to said lease, but covers and includes one-half of all the oil royalty and gas rental or royalty due to be paid *under the terms of said lease.*" (Emphasis added). The plaintiff successfully argued that his right to payments under the lease was not limited to those that accrued to the 90 acres; that the deed contained two grants, one of a reversionary right to the mineral interest in the 90 acres and another of one-half the benefits under the existing lease from the whole 320 acres. Thus, the "subject to" clause interpretation substantially affected the area covered by the deed.

In *Paddock v. Vasquez*, 122 Cal.App.2d 396, 265 P.2d 121 (1953), the two grants doctrine was applied to inconsistent fractions to affect the size of the interest conveyed. A grant of a three percent mineral interest in property subject to an oil and gas lease providing for a $^1/_8$ royalty was followed by a subject to clause that stated that the grantee was to receive $^6/_{25}$ths of all payments that might accrue under existing or future leases. Though the parties probably had erroneously completed the deed on the mistaken premise that the grantee should receive a share of the royalties under the existing lease equal to three percent of production without deduction for costs ($^6/_{25}$ths $\times$ $^1/_8$ = 3%), the court awarded the grantee $^6/_{25}$ths of the royalty under any future lease in addition to a three percent mineral interest. In the context of *Paddock v. Vasquez* the two grants doctrine is particularly appalling because the court's interpretation of the second grant substantially eclipses the grant of mineral interest.

c.   Avoiding the "Subject to" Ambiguity

The two grants doctrine results from a literal, mechanistic interpretation of mineral deed terms, often contrary to the obvious intent of the parties. It has been implicitly or explicitly rejected in several other cases on the basis that the intent of the parties was to give one provision or the other priority, or by reformation. However, because of the tendency of courts to be literal and mechanistic in interpretation of deeds, the "subject to" clause can be expected to continue to trap the unwary.

Perhaps time will see evolution of the "subject to" clause to a mere exception to the warranty; a separate assignment of lease benefits should not be necessary. Until then, great care must be taken to avoid inconsistency between the grant and the "subject to" clause language. That is done in the mineral deed form in the Appendix in what has come to be called a "Hoffman clause." It provides specifically that:

> "This sale is made subject to any rights now existing under any valid and subsisting oil and gas lease of record heretofore executed; it being understood and agreed that said Grantee shall have, receive, and enjoy the herein granted interest in and to all bonuses, rents, royalties and other benefits which may accrue under the terms of said lease *insofar as it covers the above described land* . . . ." (Emphasis added).

Such language clearly avoids the argument that the "subject to" clause makes a second grant.

2. APPORTIONMENT OF ROYALTIES

As has been discussed, transfer of property subject to an existing oil and gas lease generally transfers unaccrued lease payments. This result is reached on the basis either that the lease royalty is reserved from the lease and conveyed with the land, or that the lessee's promise to pay is a covenant that runs with the land. But what if the transfer is of a subdivided part of the leased land? How should lease benefits be apportioned? There is a split of authority as to apportionment of royalties that creates another special problem of oil and gas conveyancing.

An illustration may help. Suppose that O, the owner of fee simple absolute in both the surface and minerals of a 640 acre section leases the property to A Company. O then sells the east 320 acres to X, subject to the oil and gas lease. If at the end of the first year A Company wishes to pay delay rentals, in what proportion should the payment be made to O and X? If A Company drills a well and obtains production, how should royalties be paid? The following diagram illustrates the problem:

LEASE

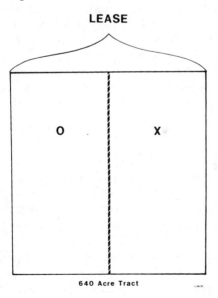

640 Acre Tract

There is no dispute over how payments of delay rentals should be made where there has been a subdivision of a leased property. Unless the grantor has retained the right to receive delay rentals explicitly (and assuming lease provisions requiring notice to the lessee of changes in ownership have been met), payments of delay rentals must be apportioned between the owners of the land; i.e., O and X will each get half. This is an application of the usual rule for rents of real property.

a. The Non-apportionment Rule

A different rule is usually applied to royalties on oil and gas production. In most states, lease royalties are not apportioned among the owners of subdivided property. Instead, the owner of the tract where the well that produces the oil and gas is located is entitled to all royalties due under the lease.

The majority non-apportionment rule is an application of the rule of capture. The reasoning of the courts is that royalties are different from rents. They are not payments that issue equally from each and every part of the land. Royalties are a right to production if and when it occurs. The rule of capture dictates that production belongs to the owner of the subdivided part upon which the producing well is located. That the subdivided tracts are subject to a single oil and gas lease does not change the result because there is nothing in typical leases inconsistent with the rule of capture. The owners of the subdivided tracts are presumed to know of the rule of capture and to intend its application. The non-apportionment rule apparently will be followed in Arkansas, Colorado, Illinois, Indiana, Kansas, Kentucky, Louisiana, Nebraska, New Mexico, Ohio, Oklahoma, Texas and West Virginia.

b. The Apportionment Rule

The minority view, the apportionment rule, treats royalties like rents. In *Wettengel v. Gormley*, 160 Pa. 559, 28 A. 934 (1894), the landmark case stating the apportionment rule, the Pennsylvania Supreme Court presumed that oil was producible equally from all parts of the subdivided land. It apportioned royalty like surface rents. The apportionment rule is followed in Pennsylvania, California and Mississippi, as well as in Ontario.

c. Understanding the Rules

Discussion of the apportionment and non-apportionment rules usually centers around which one is more "fair." In fact, either rule may be inequitable and onerous. Suppose that our example is set in an apportionment jurisdiction and the producing well is drilled on a 10 acre spacing unit in the northeast corner of the east 320 acres. An illustration follows:

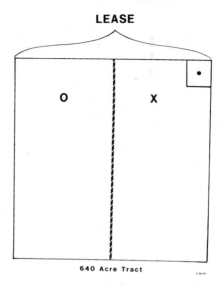

640 Acre Tract

Under such circumstances, it is highly unlikely that any of the oil or gas produced from the well comes from the west 320 acres. Apportionment is unfair. Non-apportionment seems right. But if the 640 acre lease composes a single drilling unit with the well located approximately in the middle so that the owner of the 320 acre subdivision upon which the well is not located has no possibility of a well being located on his land and drainage is a virtual certainty, the apportionment rule seems fair.

The apportionment and non-apportionment rules can best be understood as rules of property law developed to make the record title system more certain. Whenever a property subject to an oil and gas lease is subdivided into separate tracts, an ambiguity arises as to whether the grantor and grantee intended that royalties be apportioned. Adopting either the apportionment rule (by analogy to rents) or the non-apportionment rule (by applying the rule of capture) avoids the ambiguity and promotes certainty of title. Because the problem arises every time land subject to a lease is subdivided, it is more important for the legal system that there be a clearly defined rule than which rule is adopted.

d. Avoiding Conflict With the Rules

Either the apportionment or the non-apportionment rule can be a trap for the unwary or unknowledgeable. The key to avoiding problems is to be familiar with the position taken by your state and to advise clients of the potential for application of the rules.

(1) Modification by Agreement

Where the grantee's and grantor's intention to change the rule applied is clear, it will be enforced. Thus, in our example, O and X might agree to apportion royalties by a special provision in the deed transferring title to the east half of the property or in a separate agreement. Such an agreement would be binding upon O and X, though A Company could not be required to actually apportion the royalties unless it agreed to modification of its lease.

(2) Entirety Clauses

If the owner of property anticipates subsequent subdivision at the time of the grant of the lease, what some call an entirety clause may be inserted to provide for apportionment of royalties. A common formulation follows:

> If the leased premises shall [now or] hereafter be owned severally or in separate tracts, the premises nevertheless shall be developed and operated as one lease and all royalties accruing hereunder shall be treated as an entirety and shall be divided among and paid to such separate owners in the proportion that the acreage owned by each such separate owner bears to the entire leased acreage.

Entirety clauses were originally inserted in oil and gas leases by lessees when few states had adopted either the apportionment or non-apportionment rule. They were intended to clarify how royalties were to be paid. In addition, they avoided the argument that lessees were required to offset drainage from one subdivided tract to another or to install separate meters and storage facilities where wells were located on different subdivisions subject to the same lease.

Over the years, entirety clauses have fallen into disfavor. One problem has been that subdivision of the leased premises into many small tracts (e.g., as with a residential subdivision) can impose a crushing administrative burden on a lessee. Another has been that where there are several parties with various fractional interests in tracts and not all mineral interests are leased or not all leases include entirety clauses, lessees may be held liable to pay greater royalties than they had anticipated. For example, suppose that O, who owns ³/₄ mineral interest in Blackacre and ¹/₄ mineral interest in Whiteacre leases to A Company for a ¹/₈ royalty with an entirety clause, while X, who owns ¹/₄ mineral interest in Blackacre and ³/₄ mineral interest in Whiteacre, leases to A Company for ¹/₈ royalty without an entirety clause.

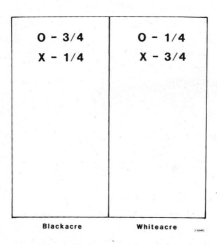

If A Company drills a successful well on Whiteacre, O may be entitled to ¹/₂ of ¹/₈ royalty by virtue of the literal terms of the entirety clause, while X will be entitled to ³/₄ of ¹/₈ royalty because his lease has no entirety clause.

As a result, entirety clauses are not often found in leases. However a variation like the following, making clear the lessee's right to operate the lease as a single lease after subdivision but without providing for apportionment of royalties, is often included:

> "If the leased premises are now or shall hereafter be owned in severalty or in separate tracts, the premises shall nevertheless be developed and operated as one Lease and there shall be no obligation on the part of the Lessee to offset wells on separate tracts into which the land covered by this Lease may hereafter be divided by sale, devise, descent, or otherwise, or to furnish separate measuring or receiving tanks."

Such provisions may be mistaken for an entirety clause.

(3) Legislative Provisions

The trap of the non-apportionment rule may also be avoided by legislative provision. In Oklahoma, the oil and gas conservation law provides specifically that all owners of property included in a spacing unit will share in $\frac{1}{8}$ royalty proportionately to their ownership of minerals. The statute apparently requires apportionment of the first $\frac{1}{8}$ royalty. It is unclear how royalty in excess of $\frac{1}{8}$ is to be handled. In this writer's opinion, the non-apportionment rule probably would be applied. Thus, in our example on page 139, if a $\frac{3}{16}$ royalty were provided by the lease, O and X would share $\frac{1}{8}$ royalty and X would take the remaining $\frac{1}{16}$.

3. TOP LEASING

A top lease is an oil and gas lease covering property already subject to an oil and gas lease; it sets on top of an existing lease. Since the mineral interest owner has already leased his mineral rights, a top lease is a partial alienation of the possibility of reverter retained by the mineral interest owner under the original or "bottom" lease.

There are two kinds of top leases. Sometimes a lessee will top lease himself to extend the duration of his rights to the property. Such a top lease is called a *two-party top lease*. Where development takes place before the bottom lease on property subject to a two-party top lease expires, there is disagreement as to which lease provisions control. Where a top lease is taken by a person other than the holder of the bottom lease, a *three-party top lease* is created. The top lessee speculates that the bottom lessee will let the bottom lease terminate so that the top lease will become possessory.

Until relatively recently, many in the oil industry regarded top leasing as immoral. Rising prices for oil and gas and increased competition for leases in the 1970's changed that perception. Most companies engaged in oil and gas development now take top leases.

a.   The Rule Against Perpetuities

A top lease may violate the rule against perpetuities. If a top lease is created by modifying the term clause of a typical lease so that it reads "This lease shall be effective from and after the termination of [description of the bottom lease]," the rule is breached. The top lessee's interest will not necessarily vest within the perpetuities period because the bottom lease may be extended by production for a time longer than the perpetuities period. Because of this problem, top leases generally are prepared either to become effective immediately or at a date certain within the perpetuities period.

b.   Obstruction

Another pitfall in preparing top leases is the equitable doctrine of obstruction. If a top lease is prepared without reference to the existence of the prior existing lease, some cases have suggested that the title of the bottom lessee is clouded, whether the top lease is recorded or not. The doctrine of obstruction suspends the running of time under the bottom lease for the duration of the obstruction or extends the primary term of the bottom lease for a reasonable period of time after its removal.

Most top leases attempt to avoid obstruction of the bottom lease by providing specifically that the top lease is subordinate to the bottom lease and subject to all of its terms and conditions. In addition, though it should not be necessary, most top leases contain a specific undertaking by the top lessor (who is also the lessor of the bottom lease) not to extend the bottom lease or to grant a new lease to the bottom lessee. These devices should be effective. However, there is little definitive precedent.

PART III

OIL AND GAS LEASING

CHAPTER 8

ESSENTIAL CLAUSES OF MODERN OIL AND GAS LEASES

A. PURPOSE OF THE LEASE

Oil and gas leases are the central documents to oil and gas development. They are structured very differently from ordinary real property leases. The key to understanding them is to identify the fundamental goals the industry has in leasing. There are two:

(1) the industry seeks the *right* to develop the leased land for an agreed term without any *obligation* to develop;

(2) if production is obtained, the industry wants the right to maintain the lease for as long as it is economically viable.

Both goals arise from the economic realities of the oil and gas business. The lessee seeks an option to develop for an agreed term because he does not know when a lease is taken whether there is petroleum under the leased land. Application of modern geological and geophysical techniques increases the odds of finding oil and gas in commercially productive quantities, but there can be no certainty until the risk of drilling a well is taken. Whether it will make sense to take that risk depends upon a variety of economic factors, including the supply and demand for oil and gas, tax structure and incentives, and applicable regulatory policies, that often cannot be assessed when a lease is taken. Therefore, the lessee's first goal is to obtain the right to operate without accepting any obligation to drill.

The second goal sought by lessees—the right to maintain a lease as long as it is profitable once production is obtained—is also motivated by economic realities. The lessee wants to maximize profit on leases upon which he has successfully taken the drilling risk. The right to maintain the lease indefinitely is essential because it is impossible to predict how long a given well or lease will produce profitably. Profitability is a function not only of the amount of oil and gas in place but also of the porosity and permeability of the structures in which they are found, the technology available to extract hydrocarbons, the supply and demand for energy, and tax and regulatory policies. Therefore, modern oil and gas leases almost always are drafted to extend for so long as there is "production in paying quantities," "capability of production in paying quantities," or "operations for oil and gas production."

In this chapter we will examine the essential clauses of modern oil and gas leases and consider the nature and extent of the rights they create.

B.   THE NATURE OF THE LEASE

A modern oil and gas lease is a unique instrument. Essentially, it transfers a mineral owner's rights to search for, develop and produce oil and gas from leased lands during the lease term. It is difficult to fit into existing legal categories. It is both a conveyance and a contract, more a deed than a lease, and it creates rights in the lessee that have proved hard to classify.

1.   BOTH A CONVEYANCE AND A CONTRACT

In most states, an oil and gas lease is both a conveyance of mineral rights from the lessor (the mineral owner) to the lessee (the oil company) and a contract between the lessor and the lessee relating to the development of minerals. It is a conveyance because the mineral owner who grants a lease transfers his rights to the property. It is a contract because the oil company that receives that transfer accepts it with certain conditions and obligations attached.

Louisiana is an exception. Under Article 114 of the Louisiana Mineral Code, a mineral lease is solely a contract under which the lessee is granted the right to explore for and produce minerals. Mineral leases are not subject to prescription for non-use. However, the Code provides that a mineral lease may not be continued for more than ten years without drilling, mining operations or production.

2.   MORE A DEED THAN A LEASE

As has been discussed in Chapter 5, oil and gas leases are different from ordinary real property leases in at least three respects: (1) the lessee has the right not only to use the land but to take substances of value from it; (2) the lessee's rights are not limited to a term of years; and (3) the lessee's rights to use the land are not exclusive but must be shared with the surface owner. Therefore, an oil and gas lease is structured more like a deed of easement or a mineral deed than a lease of real property.

3.   LEGAL CLASSIFICATION

The lessee's interest under an oil and gas lease has been classified in a variety of ways.   Some states, including Texas, have described it as an estate in fee simple determinable in the oil and gas in place.   Others, including Oklahoma, California, Montana and Wyoming have classified the lessee's interest as a *profit a prendre.*   Kansas generally treats leases as creating a license. Several cases have even classified the lessee's interest under a lease as an inchoate right that will become a vested tenancy only after production.

4.   ESSENTIAL PROVISIONS

The essential provisions of an oil and gas lease are those necessary to make a valid transfer of rights and accomplish the lessee's fundamental goals.   Generally, these can be found in just three clauses; the granting clause, the term clause, and the drilling-delay rental clause.   In the remainder of this chapter, we will consider those clauses and issues that frequently arise under them.

C.   GRANTING CLAUSE

The granting clause of an oil and gas lease spells out the rights that are granted by the mineral interest owner to the lessee.   The effect is to grant to the lessee the right to search for, develop, and produce oil and gas from the leased premises without imposing any obligation to do so.   To be valid, the granting clause must identify the size of the interest granted, the substances covered by the lease and the land covered by the lease.   In addition, most lease granting clauses specifically indicate uses permitted.

1.   SIZE OF THE INTEREST GRANTED

A peculiarity of oil and gas leases is that they are structured as if the lessor were leasing 100% of the mineral interest, even though the mineral interest may be owned concurrently.   Thus, a lessor of a $\frac{1}{16}$ fractional interest in the minerals will be presented with an oil and gas lease completed as if he owned 100% of the mineral rights.   This practice grows from the fact that leases are often acquired before a comprehensive title search is done.   In addition, mineral rights have often been split into tiny fractions, many of which have become "lost" over the years. The probability is quite high that a lessee will discover after a lease is taken that his lessor had a different portion of the mineral interest than it was thought he had when the lease was taken. As a result, lessees prefer to take leases from all mineral interest owners as if each owned 100% of the mineral interest.

2. SUBSTANCES COVERED BY THE GRANT

The purpose of an oil and gas lease is to give the lessee the right to search for, develop and produce oil and gas on the leased premises. However, oil and gas wells may produce valuable substances in addition to oil and gas. In some parts of the country, helium, carbon dioxide, or sulphur may be produced with oil and gas. Lessees generally want the right to anything of value that is produced with oil and gas.

In addition, there has been confusion as to what oil and gas are. In the early part of the century, courts in Oklahoma held that casinghead gas (gas produced with oil from oil wells) was neither oil nor gas. Therefore, a lease covering oil and gas only would not entitle the lessee to casinghead gas. More recently, it has been successfully argued that helium is gas within the meaning of "oil and gas" as used in an oil and gas lease. See *Northern Natural Gas Co. v. Grounds*, 441 F.2d 704 (10th Cir.*1971*), *cert. denied* as out of time, 404 U.S. 951 (*1971*).

[*155*]

To avoid dispute, most oil and gas leases cover more than just "oil and gas." For example, the granting clause (paragraph 1) of the Texas lease in the Appendix covers "oil and gas and all other hydrocarbons." The reference is intended to make clear that the lease applies to liquid and gaseous hydrocarbons even if they are not considered to be oil and gas. Other commonly used lease formulations define oil and gas as including "all hydrocarbons and other substances produced therewith" (which is used in Colorado lease in the Appendix) or "oil, gas and all other minerals." Because of the ambiguity of general references, specificity is desirable from the viewpoint of both the lessor and the lessee.

3. LAND COVERED BY THE LEASE: THE MOTHER HUBBARD PROBLEM

The standard for sufficiency of the description in an oil and gas lease is the same as that for other conveyances; it must be possible to locate the land. Generally, oil and gas lease descriptions use either the metes and bounds system or the rectangular system discussed in Chapter 5.

In addition to a legal description of the property covered by the lease, many leases used where property descriptions are likely to contain errors have a more general description. Often they provide that the lease is intended to cover all the land owned by the lessor in the area. A common broad formulation used is in paragraph 1 of the Texas lease form in the Appendix:

"This Lease also covers and includes any and all lands owned or claimed by the Lessor adjacent or contiguous to the land described hereinabove, whether the same be in said survey or surveys or in adjacent surveys, although not included within the boundaries of the land described above."

This is frequently called a "Mother Hubbard" clause or "cover all" clause. It is intended to protect the lessee against inaccuracies in the legal description by covering all the land owned by the lessor even if it is omitted from an erroneous legal description. Frequently, Mother Hubbard clauses will include language covering after-acquired interests in the described land as well.

Occasionally, the issue of the breadth of the Mother Hubbard clause is raised. For example, suppose that O, the owner in fee simple absolute of Blackacre, a 640 acre section, grants a lease containing Mother Hubbard language like that quoted above to A describing specifically the east 320 acres of the section. An illustration follows:

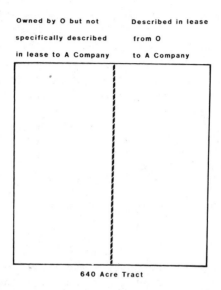

Owned by O but not
specifically described
in lease to A Company

Described in lease
from O
to A Company

640 Acre Tract

Suppose further, that A assigns the lease to B,
who has no knowledge of the specific intent of O
and A. Is B entitled to the 320 acres specifically
described or to the whole 640 acres?

If given literal effect, the Mother Hubbard language quoted would subject the full 640 acres to
the lease; the west half of the section is "adjacent or contiguous" to the east half. On the other hand, if the purpose of the language is taken
into account, the lease would be limited to the 320
acres described plus any small tracts that are
physically a part of that land but not specifically
described in the legal description. On occasion,
the courts have taken a literal view of the terms
of the Mother Hubbard clause. For example, in

Holloway's Unknown Heirs v. Whatley, 133 Tex. 608, 131 S.W.2d 89 (1939), a Mother Hubbard clause in a deed that provided that "if there is any other land owned by me in Liberty County, Texas . . . it is hereby conveyed, the intention of this instrument being to convey all land owned by me in said County," was held to cause the conveyance of ½ interest in the minerals previously reserved by the grantor. *Smith v. Allison*, 157 Tex. 220, 301 S.W.2d 608 (Tex.1956), is probably more representative of the attitude of the courts toward such clauses. There the Texas Supreme Court refused to give literal effect to a Mother Hubbard clause in a deed in circumstances similar to those of our example. However, because of the likelihood of confusion over their scope, lessors' attorneys generally view Mother Hubbard clauses with disfavor.

4. USES PERMITTED BY THE GRANT

a. General Principle: Lessee's Right to Use Burdens the Surface

An oil and gas lease gives a lessee an implied right to reasonable use of the land surface to locate, develop and produce oil and gas under the land. The courts have reasoned that since the mineral interest owner has the right to search for, develop and produce oil and gas from the premises and the purpose of the lease is to transfer that right to the lessee, the parties must have intended that the lessee acquire the right to use the surface of the land, even if that is not specifically stated. The commonly stated formulation of the law is that the lessee has an implied easement to use the surface of the land as may be reasonably necessary to obtain the minerals covered by the lease. The lessee's interest is the dominant estate. The surface of the land is servient to his right of use.

The general principle is given broad application. It gives the lessee discretion both as to the *kinds* of uses and to the *location* of those uses. Specific applications have included the right to conduct seismographic tests, to build roads and construct drilling sites, to erect oil storage tanks and power stations to power pumping units, and to conduct water flood programs to maintain production, even though use of potable ground water was required. Such applications have led to the statement that the mineral interest or leasehold interest is superior to the surface interest.

b.   Limiting Factors

The superiority of the mineral lessee's interest over the surface interest is limited by at least four countervailing principles. The lessee's use must be: (1) a reasonable use; (2) in accord with the accommodation doctrine; (3) for the benefit of the minerals under the land leased; and (4) in accord with the terms of the lease and applicable statutes, ordinances, rules and regulations.

(1) Reasonable Use

Limitation to reasonable use is inherent in the statement of general principle. The lessee has an implied easement to use the surface of the land in such ways and in such locations as may be *reasonably necessary* to obtain the minerals. The easement is exceeded if the use is not necessary or if it is unreasonable.

If the lessee's use of the land is unreasonable or unnecessary, his liability to the lessor is no different from what it would be under the same circumstances to an adjoining landowner. Thus, lessees have been held liable for damages for negligent pollution and for nuisance for failing to plug abandoned wells and remove equipment and cement foundations. Where damages will not be an adequate remedy, a lessor may obtain an injunction to prohibit unreasonable use.

Reasonableness is an important limitation on the right of surface use. It is based upon the conventions and morals of contemporary society. Thus, uses that were considered reasonable in cases considered by the courts a generation ago may no longer be permitted. For example, it is likely that a greater degree of surface damage was considered acceptable in the 1950's than will be accepted in the 1980's. Determination of what is reasonable depends upon all of the facts and circumstances. Since the facts and circumstances of each case are generally established by a jury, changes in contemporary standards of reasonableness are quickly reflected by the judicial system.

(2) The Accommodation Doctrine

The lessee's use of the land must also comply with the accommodation doctrine. The accommodation doctrine is a prime example of the responsiveness of the judicial system to changing standards of reasonableness. It was first stated as a separate principle from the reasonableness requirement by the Texas Supreme Court in *Getty Oil Co. v. Jones,* 470 S.W.2d 618 (Tex.1971). Jones, the owner of severed surface rights, sought damages from Getty Oil Company as a result of Getty's interference with his irrigation farming. Jones had drilled water wells and installed moving irrigators that were elevated approximately 8 feet off the ground and pivoted in a circle. Subsequently, Getty drilled two wells on Jones' property under authority of a lease from the severed mineral owners. Getty's wells required pumping units. The pumping units installed were substantially higher than the irrigators and the irrigators could not function. Jones contended that Getty's use was beyond the scope of its right because it effectively precluded him from farming the land. Getty asserted that its pumping units were reasonably necessary to produce the oil.

The Texas Supreme Court held in favor of Jones. It held that where a severed mineral interest owner or lessee asserts rights to use of the

surface that will preclude or impair existing surface uses, the mineral owner or lessee must accommodate the surface uses if he has reasonable
alternatives available. The court found that Getty could have sunk its pumping units below the
surface of the ground and avoided interference
with Jones' irrigators.

(A) RATIONALE OF THE ACCOMMODATION DOCTRINE

The rationale of the Texas Supreme Court was
based upon Professor Eugene Kuntz's manner of
enjoyment theory, which has been discussed (see
pages 115–116) in conjunction with the meaning
of "minerals". It reasoned that the intention
where there is a severance of mineral rights from
surface rights (or a grant of a lease) is that both
the mineral lessee and the surface owner should
have valuable estates. Therefore, an oil and gas
lessee should be required to accommodate uses of
the surface wherever possible.

(B) ELEMENTS OF THE ACCOMMODATION DOCTRINE

As articulated by the Texas Supreme Court in
Getty Oil Co. v. Jones, the accommodation principle is limited by three requirements: (1) there
must be an existing surface use; (2) the proposed
use must substantially interfere with the existing
surface use; and (3) the lessee must have reasonable alternatives available. In *Sun Oil v. Whita-*

ker, 483 S.W.2d 808 (Tex.1972), the Texas Supreme Court took an even more limited view of the accommodation doctrine. It permitted a lessee to deplete the surface owner's ground water reserves for a secondary recovery water flood project even though water could have been purchased at a modest cost from a nearby river. The court held that alternatives available to the lessee, in order to be reasonable, must be available on the leased premises.

Logically, the accommodation doctrine need not be so limited. The manner of enjoyment theory is premised upon the probable intention of the parties to the mineral severance or lease of the minerals. It is just as likely that the parties intended a balancing of the economic consequences of accommodation as that they intended that it be available only in the narrow circumstances defined by the Texas Supreme Court; their intent was general, not specific. Other courts that have considered conflicts between surface owners and mineral owners or lessees since *Getty Oil v. Jones* and *Sun Oil v. Whitaker* have recognized an obligation on the part of the mineral owner or lessee to accommodate the surface owner's interest without that limitation. As this is written, the accommodation doctrine has been specifically recognized in North Dakota, Arkansas and Utah. In this writer's opinion, it will be generally accepted, probably in a broader form than adopted by the Texas Supreme Court.

(3) For the Benefit of the Minerals Under the Surface

A third limitation to the right of the mineral interest owner or lessee to use the surface of the land is that the use must be exclusively to obtain the minerals under the land. Use of the land surface for the benefit of adjoining tracts exceeds the scope of the implied easement.

(A) APPLICATION OF THE LIMITATION

For example, suppose that O, the owner in fee simple absolute in Blackacre, granted an oil and gas lease to A Company. A Company also took an oil and gas lease on Whiteacre, an adjoining tract, from X, owner of Whiteacre in fee simple absolute. A Company now wishes to drill an exploratory well on Whiteacre. May it construct an access road across Blackacre? If the well on Whiteacre is successful, may A Company construct a pipeline across Blackacre to serve the well on Whiteacre? May it erect storage tanks on Blackacre or drill a well to dispose of waste salt water on Blackacre?

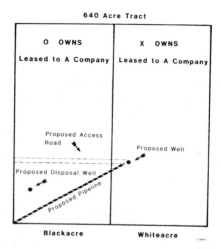

A Company could not use the surface of Blackacre for any of the uses desired. O's lease to A
Company grants A Company the right to use the
surface of O's land, Blackacre, for the benefit of
the minerals under Blackacre. Use of Blackacre
for the benefit of Whiteacre may be enjoined or
may result in award of damages.

A harder question is whether A Company has
the right to use the surface of Blackacre in conjunction with operations being conducted by A
Company on *both* Whiteacre and Blackacre. Suppose, for example, that A Company drills a well
on Blackacre and then wishes to drill a well on
Whiteacre and use the access road, pipeline, salt
water disposal well, and storage facilities it has
erected on Blackacre in conjunction with its operations on Whiteacre. Would such uses be permit-

ted? Here again, a strict view is taken of the scope of the easement granted. The lessee has no right to conjunctive uses.

The hardest question is whether A Company may use the surface of Blackacre in conjunction with Whiteacre where the properties have been compulsorily pooled or included in a spacing unit established by a state conservation agency. Where O owns both the surface and minerals, the answer is affirmative because O will benefit from the well drilled; the action of the conservation agency does not deprive O of his property, it protects his correlative rights. But what if O, the lessor, owns only the mineral rights? Can the severed surface owner, who is not benefitted or protected by the state's action, prevent use of the surface of Blackacre in conjunction with Whiteacre? He cannot because the state action is an exercise of the police power for which the surface owner is not entitled to redress. In addition, pooling protects surface owners as well as mineral owners, for without it the surface could be burdened by over-drilling.

(B) EXPANDING THE IMPLIED RIGHT

Special easements must be obtained for use of land surface in conjunction with operations on other leases. Generally, such easements are conveyed separately from the lease. But occasionally, such easements are included in the granting clause of the lease. The Colorado lease form in the Appendix is an example:

> Lessor . . . hereby grants, leases and lets exclusively to lessee the land described below for the purpose of [searching for, developing, and producing oil and gas] . . . together with all rights, privileges, and easements *useful* for lessee's operations on said land and *on land in the same field with a common Oil and Gas Reservoir.* . . .

This grant expands the scope of the easements generally implied in two ways. First, the lessee is granted the right to use the land for those purposes that are "useful" as well for those that are "reasonably necessary." Second, the lessee is given the right to use the surface for the benefit of other lands.

Such language will be interpreted narrowly by the courts. The right to use the land for purposes "useful" to the lessee might well be held to be nothing more than a restatement of the reasonable use principle. The grant of the right to use the land surface in conjunction with other lands with a common oil and gas reservoir may be held to require that there be producing operations upon Blackacre or that Blackacre be pooled with the property upon which producing operations are taking place. Furthermore, the grant of a specific easement by lease is inherently limited where the lease is from a severed mineral interest owner. Where the lessor owns the fee simple interest in the property he holds all of the "bundle" of rights. He may grant specifically to the mineral lessee broader rights than those usually implied. However, where the lessor is the holder of a severed mineral interest, he does not have rights of surface use beyond those normally implied and so cannot grant them to a mineral lessee.

(4) In Accordance With Lease Terms and Applicable Statutes, Ordinances, Rules and Regulations

(A) RESTRICTION BY LEASE PROVISIONS

Modern oil and gas leases frequently contain specific provisions limiting the implied easement for surface use. For example, paragraph 7 of the Texas lease form in the Appendix provides that the lessee must pay damages for growing crops and bury pipelines installed on the leased premises. The principle of reasonable use would probably require neither. The commonly encountered lease clause clarifies the intention of the parties and limits the right of surface use of the lessee.

The leasing boom of the late 70's and early 80's gave lessors the economic leverage to demand much more extensive and onerous provisions. It is common to see page after page of surface use limitations and surface use procedures in addenda to leases on large properties. It is important from lessor's viewpoint that surface use limitations be agreed to before or contemporaneously with the grant of the lease; otherwise they may fail for lack of consideration. Also, to be binding upon third parties they must be reflected on the record, either in an addendum to the lease or (in many states) by reference to an unrecorded agreement.

(B) RESTRICTION BY STATUTES, ORDINANCES, RULES, OR REGULATIONS

Restrictions on the implied right of surface use by statutes, ordinances, rules and regulations, are a fast-growing phenomena. Requirements that lessees obtain drilling permits and properly plug and abandon wells have been common since the early part of the 20th century. In the 1960's and 1970's, many states set standards for clean-up and restoration of the surface after drilling operations and upon abandonment of the lease. State and federal governments have generally used rules and regulations to impose many of the same kinds of surface use restrictions that large private mineral interest owners have defined in oil and gas leases. Generally, such standards have been held valid as legislative or regulatory definitions of reasonable use.

However, the most recent type of legislative restriction on surface use goes beyond defining reasonable use. Surface damages acts, enacted in several oil producing states in the late 1970's and early 1980's, require payment of damages for use of land surface in oil and gas operations. They reverse the traditional principle of law that the mineral interest owner or oil and gas lessee is entitled to reasonable use of the land surface to obtain minerals without the landowner's permission and without payment.

The roots of surface damages acts are in the practice of the oil industry. To avoid argument whether a particular use is reasonable, most oil companies pay voluntary "site damages" to surface owners. Landowners have come to regard these payments as their right. Disturbed by what they have regarded as inadequate damage offers and by occasional abuses of the right of reasonable use, they have pressed surface damages legislation.

There is substantial doubt that surface damages acts are constitutional. The United States Constitution provides that private property rights may not be taken without reasonable compensation and due process. Since severed mineral owners and holders of leases had the right to use the surface of the land without compensation prior to the surface damages laws, it will be argued that their rights cannot be taken by requiring damages for reasonable surface use. On the other hand, surface damages statutes may be valid if they are only of prospective application. They may also be interpreted to require payment of damages only where surface use is unreasonable. Or, a tenuous distinction may be drawn on the basis that they preserve lessees' right of use but impose reasonable cost burdens.

D. THE TERM CLAUSE

The term clause of an oil and gas lease sets the period of time that the rights given in the granting clause will extend. Typically, modern lease term clauses provide for a primary term and a secondary term. The *primary term* of an oil and gas lease is a fixed term of years during which the lessee has the right, without any obligation, to operate on the premises. The *secondary term* is the extended period of time for which rights are granted to the lessee once production is obtained. A common formulation is:

"[T]his Lease shall be for a term of—years from this date, called 'primary term,' and as long thereafter as oil or gas are produced . . . ."

The term clause is the explicit statement of the two key goals sought by lessees in oil and gas leases, which have been discussed at pages 150–151.

1. THE PRIMARY TERM

The purpose of the primary term of a lease is to give the lessee adequate time to acquire additional leases in the area, to do geological and geophysical tests to evaluate whether to drill a test well, and to arrange for financing and support services to drill. The length of the primary term is established by negotiation. It is a function of what the market will bear. It is determined by the bargaining leverage of the parties and the amount of the bonus that the lessee is willing to pay. It may be as long (except in Louisiana and Tennessee where it may not exceed ten years) or short as the parties agree. Ten years was once a common primary term. It is still frequently seen in leases in unproven and marginally producing areas. Terms from one to five years are more typical in states with established oil and gas production.

The primary term states the maximum period of time for which the lessee can maintain the lease rights without drilling. It may be cut short either by surrender of the lease by the lessee or by failure to pay delay rentals properly. It may be extended by production or, under many leases, by operations to the secondary term.

2. SECONDARY TERM

The purpose of the secondary term is to give the lessee the right to hold a producing lease as long as it is economically viable to do so. The secondary term is an indefinite period of time because it is impossible to determine for how long a lease will be profitable at the time it is granted. However, because the secondary term is not specifically designated, disputes frequently arise. Among them are: a) what is "production" necessary to extend the lease; and b) when does production cease and the lease terminate.

a. The Meaning of "Production"

Except in a few states, actual production is required to extend an oil and gas lease to the secon-

dary term. The lease terminates automatically at
the end of the primary term without it. The rationale for this view is that the business deal
struck by the parties is expressed in the term
clause, which says that the term will extend as
long as there is "production." The literal wording of that clause must be given effect; without
actual production, the lease terminates automatically at the end of the primary term, unless some
other provision changes that result.

The majority position implicitly requires marketing as well as production. The economic basis
of the transaction is the rationale. There is no
purpose of production if it cannot be marketed.

The classic case illustrating the strictness of
the majority view is *Baldwin v. Blue Stem Oil
Co.*, 106 Kan. 848, 189 P. 920 (1920). There, the
lessee had oil and gas leases with a primary term
that ended January 17, 1919. There were no provisions in the leases for extension by operations.
Blue Stem Oil Company did not begin drilling operations until December 7, 1918. It did not complete them until after the end of the primary
term. It attempted to excuse its failure to attain
actual production by asserting as affirmative de-
fenses that it had been plagued with troubles
such as an inadequate supply of water, flooding,
blizzards, shortages of coal, personnel and equipment, and adverse governmental regulation. Despite this impressive listing, the Kansas Supreme
Court affirmed judgment on the pleadings for the
lessor on the grounds that the leases had expired

by their own terms. Actual production of oil and gas is required to extend a lease into its secondary term.

There is a respectable minority view, of which Oklahoma and West Virginia are the main proponents, that an oil and gas lease will not terminate if there is a discovery of oil or gas prior to the end of the primary term; actual production is not necessary to preserve the lease, though "discovery" probably requires completion and capability of production. There are cases in Montana, Wyoming, Kentucky and Tennessee suggesting that the discovery of gas will be sufficient to extend the lease to its secondary term, but actual production of oil will be required to extend the lease. The rationale of such a distinction is that oil can be produced and stored without actual marketing, while gas cannot be economically stored.

Both minority views interpret "production" in the term clause by reference to the essential purposes of the lease rather than to the literal language used. The major purpose of a lessor in granting an oil and gas lease is to obtain development of his property. That purpose has been substantially served when there is a well capable of production.

b. How Much Production Is Required: "Production in Paying Quantities"

A literal construction of "production" in the term clause of an oil and gas lease would mean that small amounts of production would suffice to extend the lease indefinitely. However, with the exception of scattered cases, the courts that have considered the issue have concluded that the production must be "in paying quantities to the lessee."

The rationale for requiring production "in paying quantities" to extend the lease to its secondary term is convincing. Modern oil and gas leases have an indefinite secondary term to avoid the problem of termination at the end of some arbitrarily fixed term while it is still profitable to produce the leased property. From the viewpoint of both lessees and lessors, the lease is an economic transaction. When it no longer is profitable, it should terminate. Otherwise, lessees will be permitted to speculate with lessors' interests.

Logical though the "in paying quantities" standard is, it has proved difficult to apply in practice. Perhaps the best statement of the meaning of the standard is found in *Clifton v. Koontz*, 160 Tex. 82, 325 S.W.2d 684 (1959):

> "[T]he standard by which paying quantities is determined is whether or not under all the relevant circumstances a reasonably prudent operator would, for the purpose of making a profit and not merely for speculation, continue to operate a well. . . ."

> . . .

> "In determining paying quantities, in accordance with the above standard, the trial court

must necessarily take into consideration all matters which would influence a reasonable and prudent operator." 325 S.W.2d at 691.

Note that the standard does not require that the lessee have a reasonable expectation of recovering his costs of drilling and completing the wells on the lease. Once a well is put into production, it will make sense to continue to operate if operating revenues will be greater than operating costs, even though the costs of drilling and completing will never be recovered. The reasonable prudent operator must make business decisions based upon the facts as they are. The issue is whether continued operations will produce more revenue for the lessee than they cost over a reasonable period.

Application of the standard is difficult. It requires consideration of at least three factors: (1) what revenues and expenses are to be taken into account in determining paying quantities, (2) over what period of time the calculation should be made, and (3) what other circumstances should be taken into account.

(1) Determining Operating Revenues and Operating Costs

(A) OPERATING REVENUES

Determining operating revenues is relatively easy. All revenues from the sale of production are taken into account in determining paying quantities. Payments that reflect a return of capital rather than operating income (e.g., sale of an oil storage tank no longer needed) are not taken into account. Likewise, the amount of the revenues paid to the mineral interest owners as royalty under the lease is excluded; the landowner's royalty is not revenue *to the lessee.*

There is a conceptual problem with treatment of revenues subject to claim by overriding royalty interest owners. As a general rule, the share of revenues due to overriding royalty interests is taken into account in determining paying quantities; only the landowner's royalty is excluded. The rationale of this distinction may be difficult to understand, for revenues that are paid to overriding royalty interests add nothing to the lessee's profitability. But most overriding royalty interests are held by persons who receive them as compensation for their help in structuring ventures. Payments of overriding royalty enhance the lessee's profitability indirectly; without them, his direct costs would be higher. Equity and public policy may also be a factor. Determination of "paying quantities" is inherently imprecise. Fairness demands that a lessee who has obtained production be given benefit of the doubt. So does the public policy to maximize recovery of oil and gas.

(B) OPERATING COSTS

There is much more uncertainty about what kinds of expenses are taken into account in determining "paying quantities." Direct operating expenses such as the wages of the employees who service the well, the cost of electricity to run pumping units, and day-to-day maintenance, are operating expenses to be taken into account. However, there is substantial dispute whether depreciation and administrative overhead costs should be considered.

Logically, direct depreciation and administrative expenses should be weighed as operating costs in the paying quantities equation. If there is a pumping unit on a well site that is worth $10,000 this month but will be worth only $9,500 next month, the $500 loss in the salvage value of the pumping unit is a real expense factor which the economically-oriented lessee will take into account. Likewise, if abandoning an oil and gas lease would permit closing of a district production office or laying off supervisory personnel, expenses of maintaining those services are logically costs of operating the lease.

Practical considerations may outweigh logical symmetry, however. In 1979, the Oklahoma Supreme Court decided in *Stewart v. Amerada Hess Corp.*, 604 P.2d 854 (Okl.1979), that depreciation of lifting equipment must be considered an expense in determining paying quantities. That case has been followed by a flurry of litigation over what equipment constitutes "lifting equipment" and how lifting equipment should be depreciated. The equitable and public policy principles noted in discussion of operating revenue also dictate against consideration of such costs. Perhaps for these reasons, the Supreme Court of Kansas in *Texaco Inc. v. Fox*, 228 Kan. 589, 618 P.2d 844 (1980), specifically rejected the Oklahoma position, and the Texas cases suggest that while such costs may be taken into account, they are not required to be considered.

(2) The Time Factor

The time period over which operating revenues and costs are to be considered is of equal importance to what revenue and expenses are to be taken into account in determining paying quantities. This issue is capable of even less precise definition than what constitutes operating revenues and costs.

Before a lease will be found to have terminated, there must be history of nonpaying production sufficiently long to suggest that the lessee's continued operations are speculative. Few businesses and few oil and gas leases will operate at a profit all the time. Profitability of oil and gas operations is particularly sensitive to seasonal transportation difficulties and changes in market demand. Therefore, production in paying quantities does not cease the first day or the first month that the lease fails to operate profitably. Production ceases only when the lease is losing money and there is no reasonable expectation that it will become profitable.

The cases on the issue are consistent only in that all select at least a year as the basis for determination of production in paying quantities. Many consider operating revenues and operating costs over substantially longer periods; eighteen months to three years is common. It is left to the discretion of the court to select a time period that will permit fair assessment of the potential for profitability of the lease in question.

(3) Equitable Considerations

Equitable factors are frequently taken into account by the courts in determining the appropriate time over which to consider the facts of operating revenues and operating costs. Thus, where political or economic conditions are turbulent, a longer history of operating costs and revenues will be considered than will otherwise be the case. However, in most states, once the determination of the appropriate time period has been made, the determination of whether or not the lease is producing "in paying quantities" is a matter of economic analysis. It is submitted that equitable considerations should be given equal weight with the conventional analysis. Figures as to production revenues and costs are only part of the circumstances that determine an operators prudence in continuing operations.

E. THE DRILLING–DELAY RENTAL CLAUSE

The purpose of the delay rental clause is to ensure that the lessee has no obligation to drill dur-

ing the primary term by negating any implied obligation to test the premises. Before drilling-
delay rental clauses became common in oil and
gas leases, many courts held that lessees had an
implied duty to drill a test well on the leased
premises within a reasonable time after grant of
the lease. The rationale for the implied covenant
was that the major consideration for the grant of
the lease by the lessor was the expectation that
the property would be tested within a reasonable
time. The courts' determination of what was a
reasonable time ranged from a few months to
several years, depending upon the circumstances.
Lessees could not rely upon a long stated term
alone to preserve their rights.

Except in a few states, where a drilling delay
rental clause is included in a lease there is no implied covenant to drill a test well. For example,
in *Warm Springs Development Co. v. McAulay*,
576 P.2d 1120 (Nev.S.Ct.1978), the court refused
to imply a covenant to test in a geothermal lease
for a primary term of 20 years without even considering the bonus paid, even though the delay
rentals provided for were only 10¢ per acre per
year. The specific right to hold the lease during
the primary term by payment of delay rentals dis-
claims the implied covenant to test.

Lessors do not generally resist drilling-delay
rental clauses. In part, this is because the
clauses have become customary in lease forms.
Lessors commonly enter into leases with the expectation that development will not occur, if ever,
until close to the end of the primary term. Those

who consider the timing of the drilling of the first well important will negotiate for a short primary term or for a specific drilling obligation. Moreover, many lessors look forward to periodic receipt of delay rental payments.

From the viewpoint of the lessee, it would be ideal if the option period of the lease primary term could be extended indefinitely. Early in the century, leases giving the lessee the right to extend the term of the lease indefinitely without production were commonly seen. They were called "no term" leases because they could be extended indefinitely by payment of delay rentals. No term leases are in great disfavor in modern times, at least in oil and gas development. Many courts have refused to enforce them on the grounds that they create a mere estate at will, terminable by either the lessor or the lessee. Others have upheld no term leases, but with the stipulation that the lessee has an obligation to develop or release the lease within a reasonable time. However, the primary reason that no term leases have fallen into disuse is that they are not acceptable in the market place; both mineral owners and lessees demand more certainty than they provide.

1. "UNLESS" v. "OR" LEASES

a. The "Unless" Lease

There are two polar types of drilling-delay rental clauses in current use. The more common
[*183*]

type in all of the states except California, is the "unless" clause. It is structured so that it automatically terminates the lease *unless* a well is commenced or delay rentals are paid prior to the date specified. Paragraph 5 of the Texas lease in the Appendix is typical:

> "If operations for drilling are not commenced on said land, on an acreage pooled therewith as above provided for, on or before one year from the date hereof, the Lease shall terminate as to both parties, unless on or before such anniversary date Lessee shall pay or tender to Lessor, or to the credit of Lessor in the _____ Bank at _____, Texas, (which bank and its successors shall be Lessor's agent and shall continue as the depository for all rentals payable hereunder regardless of changes in ownership of said land or the rentals) the sum of ($_____), herein called rentals, which shall cover the privilege of deferring commencement of drilling operations for a period of twelve months. In like manner and upon like payment or tender annually, the commencement of drilling operations may be further deferred for successive periods of twelve months each during the primary term hereof. . . ."

The clause creates a *special limitation* on the primary term of the lease; it modifies the term clause of the lease by making periodic commencement of drilling operations or payment of delay rentals essential to hold the lease during the primary term. Under an "unless" drilling-delay rental clause, termination occurs automatically at the end of the annual period unless drilling operations are begun or delay rental is paid.

b. The "Or" Lease

The less common but increasingly popular "or" drilling-delay rental clause does not cause the lease to terminate automatically if the lessee fails to commence drilling operations or pay rentals in a timely fashion. In contrast to the "unless" clause, it imposes an affirmative duty upon the lessee to pay the delay rentals:

> "Commencing with the first day of the second year of the term hereof, if the lessee has not theretofore commenced drilling operations on said land or terminated this lease as herein provided [by surrender], the lessee shall pay or tender to lessor annually, in advance as rental, the sum of $1 per acre per year for so much of said land as may then still be held under this lease, until drilling operations are commenced or this lease is terminated as herein provided."

Under the terms of this clause, the lessee must either commence drilling *or* pay rentals *or* surrender the lease prior to the due date. Because the clause affirmatively obligates the lessee to do one of the alternatives, it is commonly referred to as an "or" drilling-delay rental clause. "Or" clauses are the rule rather than the exception in California.

c. Forfeiture of "Or" Leases

The lessor's remedy for a lessee's failure to pay delay rentals under a lease with an "or" drilling-delay rental clause is limited to a suit for the amount of the rental payment due, in the absence of special provision in the lease. A good example of this principle is *Girolami v. Peoples Natural Gas Co.*, 365 Pa. 455, 76 A.2d 375 (1950). There, after making periodic payments for eleven years under an "or" lease that required quarterly payments of delay rentals, the lessee learned of potential problems and suspended delay rental payments for two years. When payments were resumed, the lessor claimed that the lease had terminated or had been forfeited. The Supreme Court of Pennsylvania held that since the lease did not provide for automatic termination or expressly reserve the power of forfeiture to the lessor, the lessor's only remedies were action in law for recovery of the rentals or action for recision of the lease on a theory of abandonment.

Most modern "or" leases contain a forfeiture clause. The forfeiture clause gives the lessor the alternative of following its procedures to declare the lease forfeited if delay rentals are not timely paid. Usually, forfeiture clauses require notice to the lessee and an opportunity for corrective action before forfeiture can be declared. However, neither notice nor a chance to correct a failure is required. For example, in *Alexander v. Oates*, 100 Cal.App.2d 266, 223 P.2d 264 (1950), a California court had before it a commonly used California "or" form with a twenty year primary term. Its drilling-delay rental terms provided in relevant part:

"5. Commencing with the Sept. 1, 1945 [sic], . . . if the Lessee has not theretofore commenced drilling operations on said land or terminated this lease as herein provided, the Lessee shall pay or tender to the Lessor semiannually in advance as rental, the sum of One ($1.00) dollars per acre per year for the 1st 18 mo. and 2.00 per acre per yr. for the bal. of 3 yr. period [sic], for so much of said land as may then be held under this lease, until drilling operations are commenced or this lease terminated as herein provided.

"6. The Lessee agrees to commence drilling operations on said land within three (3) years from the date hereof . . . . *The Lessee may elect not to commence or prosecute the drilling of a well on said land as above provided, and thereupon this lease shall terminate.* (Emphasis added.)

. . .

"21. *Upon the violation of any of the terms or conditions of this lease by the Lessee and the failure to begin to remedy the same within 90 days after written notice from the Lessor so to do, then, at the option of the Lessor, this lease shall forthwith cease and terminate* . . . ." (Emphasis added.)

[*187*]

The lessee did not commence drilling operations within the three year period provided in paragraph 6. The lessor contended that the lease had terminated as a result. The lessee disagreed, relying upon the 90 day notice provided for in paragraph 21. The court held that the situation was governed by paragraph 6, and that notice was not required under paragraph 21.

In effect, the lease in *Alexander v. Oates* was an "or" lease during the first three years of its primary term, requiring the lessor to take affirmative action to terminate the lease by giving the lessee notice of the failure to pay delay rentals and time to correct his oversight. However, after the initial three year period, the lease became an "unless" lease that could not be satisfied by payment of delay rentals. It could be maintained for the remainder of its primary term only by periodic drilling. It automatically terminated for failure to begin drilling operations.

"Unless" drilling-delay rental clauses and "or" drilling-delay rental clauses are polar types. Under the "unless" form, the terms of the drilling-delay rental clause is a special limitation upon the primary term of the lease that will cut it short if not observed. When the "or" form is used, compliance with the terms of the drilling-delay rental clause is a covenant, and the lessor's remedy for failure to pay is a suit for damages unless forfeiture procedures are provided. Despite some judicial suggestions to the contrary, there is no reason that the parties to a lease should not be able to fashion a hybrid formulation if that more closely meets their goals.

Because the "unless" drilling-delay rental clause is the more commonly used and because most "or" leases contain forfeiture clauses, the drilling-delay rental clause in modern oil and gas leases is a frequent source of dispute and litigation. Typically, problems center upon whether the provisions of the drilling-delay rental clause have been met timely.

2. THE DRILLING OPTION

The lessee who seeks to avoid termination of his lease by compliance with the drilling-delay rental clause is presented with a choice. He can either drill or pay delay rentals within the period specified. If he chooses to drill, the problem that most often arises is dispute whether he has done so in a timely manner. There are three aspects of compliance with the drilling option: (a) the precise language used in the lease, (b) the good faith of the lessee, and (c) the lessee's due diligence.

a. Commencement v. Completion

Most modern oil and gas lease forms require that the lessee merely "commence operations for drilling" or "commence drilling operations" before the anniversary date to preserve his rights. As interpreted by most courts, the lessee has complied if he begins preliminary actions usually associated with actual drilling on the premises, in good faith, and diligently pursues them to completion. Though compliance is a question of fact, virtually any kind of work at the well site prior to the end of the primary term will likely be considered sufficient. Courts have held that digging a slush pit on the last day of the term, staking a location, delivering material, or erecting a derrick and beginning to drill a water well were sufficient to constitute commencement of operations for drilling. The cases show a clear tendency toward liberality. Any action by the lessee on the premises that shows a clear intention to develop the land will be enough to comply, as long as it is diligently pursued. It is arguable that preliminary actions that do not take place on the land, such as signing a binding drilling contract, should also qualify under that standard.

The language of the lease controls. Reference to "completion" in *Baldwin v. Blue Stem Oil Co.*, 106 Kan. 848, 189 P. 920 (1920), was held to require completion of a well rather than mere commencement. In Montana, a court has made a distinction between "commencement of operations for drilling" and "commencement of drilling operations," holding that the latter term requires actual spudding. But as a general rule, courts seem to be willing to take a liberal view of what will comply with the drilling option of the drilling-delay rental clause. Their position encourages development and avoids unfairness to lessees who have committed substantial amounts of money to development.

b. In Good Faith

The second aspect of the drilling option is that whatever preliminary work is relied upon to comply with the drilling-delay rental clause must be undertaken in good faith. Substance prevails over form. Actions which ordinarily would be considered commencement of drilling operations will be held insufficient if the courts decide that the work was a sham or that there was no intent to complete the well. For example, the Michigan Supreme Court in *Globe v. Goff*, 327 Mich. 549, 42 N.W.2d 845 (1950), held that a well had not been "commenced" where a statutorily required permit had not been acquired and no drilling contract had been executed before the end of the primary term, through the lessee had done substantial site work; the court inferred from the facts before it that the lessee had delayed committing to develop the property until it had had an opportunity to review drilling information from an offset well.

No particular action or failure by the lessee short of spudding a well is determinative. Deciding whether operations were commenced in good faith is an application of gastronomic jurisprudence to give effect to the bargain the parties to the lease made.

c. With Due Diligence

The third aspect of complying with the drilling option of drilling-delay rental clause is that once commenced, operations must be pursued with due diligence until the well is completed and put into production or plugged and abandoned. Due diligence is determined by the circumstances. The lack of precision of the standard gives the courts flexibility to protect lessors from overreaching lessees. For example, a lessee who commences drilling operations in a timely manner but removes the equipment from the premises a week after the anniversary date because of a lack of funds to pay the drilling crew probably will lose his lease because the aborted drilling operations are of no benefit to the lessor and raise an inference of lack of diligence or bad faith on the part of the operator. On the other hand, the result should be different if the equipment removal came as a result of a bona fide contractual dispute between the lessee and the drilling crew. Likewise, where shortages of equipment and crews make it difficult to obtain a drilling rig, substantial periods of time between commencement of preliminary operations on the lease and actual drilling may be permissible.

In sum, under most modern oil and gas leases with drilling-delay rental clauses drafted in terms of "commencement," the drilling option may be satisfied by substantial performance. If the lessee begins operations on the premises that are directly related to actual drilling with a good faith intention to complete a well, he will be permitted to continue his operations with due diligence until a well is completed.

3. PAYMENT OF DELAY RENTALS

Substantial performance is not sufficient to satisfy the lessee's alternative option of paying delay rentals. The option to pay delay rentals rather than commence drilling operations generally requires perfect compliance. It is often said that delay rentals must be paid (a) in the proper amount, (b) on or before the due date, and (c) to the proper parties.

The amount of the delay rental payment varies substantially from region to region. Generally, except where competition is fierce, the payments called for are nominal, running from $1 to $10 per acre per year. The amount of delay rentals is negotiable, however. Where leases are in demand they may be substantial. For example, in the Tuscaloosa trend in southern Louisiana, delay rental payments of 100% of the bonus paid are common. With bonuses ranging from $500 to $3,500 per acre there, delay rental payments are a significant factor.

However nominal the amount of the delay rental provided for, failure to pay properly generally causes automatic termination of a lease with an "unless" drilling-delay rental clause. Thus, in *Phillips Petroleum Co. v. Curtis*, 182 F.2d 122 (10th Cir.1950), a valuable lease was held to have terminated where delay rentals were not paid because a clerical employee mistakenly concluded that it was held by production. Similarly, in

Greer v. Stanolind Oil and Gas Co., 200 F.2d 920 (10th Cir.1952), the lease was 'ost where the lessee made a good faith mistake as to the due date of the delay rental payment. Underpayment or payment to the wrong person is just as fatal to the lessee as nonpayment. In *Young v. Jones*, 222 S.W. 691 (Tex.Civ.App.1920), the lease was held to have terminated where a lessee's tender was $2.96 short. The lease from the underpaid owner terminated in *Atlantic Refining Co. v. Shell*, 217 La. 576, 46 So.2d 907 (1950), when the lessee paid delay rentals in the wrong proportions to the cotenants.

A stricter standard is applied to the provisions of the "unless" drilling-delay rental clause relating to payment of delay rentals than to the provisions for commencing drilling operations. There may be three reasons for that. First, equities are more with the lessee who seeks the right to spend hundreds of thousands or millions of dollars in drilling than with one who seeks to extend his option to hold the lease by a nominal payment. Second, courts have a deep-seated abhorrence of option agreements and strictly interpret attempts to extend an option. Third, and probably most important, the provisions of the delay rental option of the lease are much less open to interpretation than the drilling option. A requirement that the lessee "commence operations for drilling" is susceptible to broader interpretation than one that the lessee "pay or tender" rentals by a specified date.

a. Protection of the Lessee by the Courts

(1) General Rule: Equity Not Applicable

Ordinarily, equity will not protect the lessee against termination of a lease containing an "unless" drilling-delay rental clause for failure to pay rentals properly. Estoppel and waiver do not apply to termination of an interest subject to a special limitation. Estoppel and waiver are concepts used to tie the hands of persons who have legal rights in situations in which it is considered unfair that they should exercise them. Termination of the lessee's interest under a lease with an "unless" drilling-delay rental clause requires no exercise of rights by the lessor. The lessee's interest lasts only so long as the terms of the drilling-delay rental clause are met; if delay rentals are not paid the interest terminates by operation of law.

(2) Revivor of the Lease

Revivor is more logically satisfying than estoppel or waiver as a theory to preserve the "unless" lease where the lessee has made an improper payment of delay rentals that has been accepted. Where the lessee has tendered the delay rental payment incorrectly, the lease automatically terminates. If the lessor accepts the payment tendered, understanding that the lease has terminated, it may be argued that his action has "revived"

the lease, if there are actions or a writing sufficient to satisfy the Statute of Frauds. However,
revivor of a lease is rarely found by the courts.

(3) Estoppel and Waiver

Notwithstanding the legal theory, there are
cases in most jurisdictions with substantial oil
and gas production that invoke equitable principles to maintain "unless" oil and gas leases
where there has been a failure to pay delay rentals properly. Many reach their pleasing but illogical results with little reasoning.

The cases that preserve leases after a failure to
pay delay rentals may be divided into at least
four classes:

(1) those in which the lessor causes late payment or underpayment;

(2) those in which payments are inadequate or
incorrect and the lessor knowingly delays
notifying the lessee of his mistake until the
anniversary date is past;

(3) those in which a late payment is accepted by
the lessor; and

(4) those in which the failure is due to failure
of an independent third party, usually the
Post Office.

The first two of the four classes are illustrated
by *Humble Oil & Refining Co. v. Harrison,* 146
Tex. 216, 205 S.W.2d 355 (1947). There the lessee
was provided by Harrison with a copy of an am-

biguous mineral interest deed to support his
claim to receive a portion of delay rentals due under a lease. Humble misinterpreted the lease.
Incorrect payments for delay rentals due were
deposited March 1 and May 8 to Harrison's credit
and acknowledged by Harrison's bank. Harrison
took no action to disavow the bank's acceptance
of the rentals until June 10. When Humble sued
to quiet title, the Texas Supreme Court held that
Harrison was estopped to assert that Humble's
lease had terminated. The basis of the decision
was that Harrison, as a party to the ambiguous
deed, had a duty to notify Humble of its mistake
in payment so that Humble could correct it.

Cases in which acceptance of late delay rental
payments is held to preserve the lease are closely
related to *Humble v. Harrison*. Both are premised upon the contractual principle of cooperation. Parties to a contract are required to act as
if they intended to make the contract work.
Therefore, where one party has made an obvious
mistake, such as underpayment of delay rentals,
the other is required to notify him promptly of
his mistake so that he can correct it, or face estoppel for his delay. By similar reasoning, the
lessor who receives late payment of delay rentals
is presumed to know that the payment is tendered with the expectation that it will maintain
the lease. The principle of cooperation will not
permit the lessor to hold the delay rentals tendered while speculating whether it would be more
profitable to declare the lease terminated or to
keep the rentals and honor the lease.

Cases refusing to find termination of leases containing an "unless" drilling-delay rental clause where there has been a failure of independent third parties turn on a different rationale. In addition to equitable considerations, those cases generally find an implied agreement that use of the services of the third party is appropriate. Most of the cases involve failure of the Post Office to deliver properly addressed and stamped letters. Closely related, however, are those involving failure of banks either to credit deposits properly to lessors' accounts or to honor lessees' checks. In either situation the leases may be preserved.

Equity is a flimsy thread upon which to hang a claim to protection for failure to pay delay rentals properly. Though equity may prevent lease termination, most reported cases hold that equitable considerations are not relevant in the circumstances presented. What circumstances will be deemed appropriate for equity to preserve the lease are so uncertain that equity's protection is of dubious value. As a result, most modern oil and gas leases provide specifically for commonly occurring problems of delay rental payments.

b.   Protection of the Lessee by Lease Clauses

(1) Payment to an Agent

Some payment problems are anticipated in the language of the delay rental clause itself. For example, in the clause quoted at page 184, the lessee is specifically permitted to make delay rental payments to a designated depository bank. This provision is more administratively feasible and more certain than direct payment to the lessor. It provides a record of the payment for the lessee, who may be making hundreds or thousands of such payments. Furthermore, a typical clause continues:

> "[P]ayment or tender of rental . . . may be made by check or draft of lessee mailed or delivered to Lessor, or to said Bank on or before the date of payment. If such Bank, or any successor Bank, should fail, liquidate or be succeeded by another Bank, or for any reason fail or refuse to accept rental, Lessee shall not be held in default for failure to make such payment or tender of rental until thirty (30) days after Lessor shall deliver to Lessee a proper recordable instrument, naming another Bank as agent to receive such payments or tenders . . . ."

The provision that payment is effective upon mailing of a check or draft is typical. Sometimes courts have been willing to imply the rights to mail and to pay by check or draft, but few lessees are willing to rely upon implication. The final part of the provision, protecting the lessee in the event of default by the depository bank, is also probably strictly unnecessary, but deemed expedient by lease draftsmen.

(2) Notice of Assignment Provisions

Most leases also contain a *notice of assignment clause* to avoid disputes over the effect of an assignment upon delay rental payments. The notice of assignment clause provides that the lessee may rely upon its records in making delay rental payments. Without such a clause there is a risk that the lessee will be obligated to review the public property records each year before the anniversary date and then to interpret any conveyances found in the record to determine who should be paid delay rentals and in what proportions they should be paid. Typical language is found in the Texas lease in the Appendix:

> "The rights of each party hereunder may be assigned in whole or in part, and the provisions hereof shall extend to their heirs, successors and assigns, but . . . no change or division in such ownership shall be binding on Lessee until thirty (30) days after Lessee shall have been furnished with a certified copy of recorded instrument or instruments evidencing such change of ownership. . . . "

Similar language was a factor in the decision of the court in *Humble Oil & Refining Co. v. Harrison* that equity prevented the lease from terminating where the lessor had failed to notify the lessee of a mistake in payment of delay rentals. An assignment provision was held in *Gulf Refining Co. v. Shatford,* 159 F.2d 231 (5th Cir.1947), to protect the lessee even where the notice was received prior to the due date of the delay rental payment, but after the payment had actually been made.

Related problems arise when the lessor dies, or when two or more persons are entitled to delay rentals. It is common for oil and gas leases to provide specifically for the payment of delay rentals in such situations. Again, the Texas form attached is a good example. Its notice of assignment clause continues:

"In the event of the death of any person entitled to rentals hereunder, Lessee may pay or tender such rentals to the credit of the deceased, or the estate of the deceased, until such time as Lessee has been furnished with proper evidence of the appointment and qualifications of an executor or an administrator of the estate, or if there be none, then until Lessee is furnished satisfactory evidence as to the heirs or devisees of the deceased, and that all debts to the estate have been paid. If at any time two or more persons become entitled to participate in the rental payable hereunder, Lessee may pay or tender such rental jointly to such persons, or to their joint credit in the depository named herein; or, at the lessee's election, the portion or part of said rental to which each participant is entitled may be paid or tendered to him separately or to his separate credit in said depository; and payment or tender to any participant of his portion of the rentals hereun-

[*201*]

der shall maintain this Lease as to such participant."

Elaborate notice of assignment clause provisions rarely answer all questions, however. For example, what if the lessee considered the certified copy of a transfer of interest provided to be ambiguous? Could the lessee ignore it and pay delay rentals as provided in the drilling-delay rental clause of the lease? The language of the assignment clause does not address the issue. However, the answer is probably in the affirmative *if* the courts find, with the benefit of hindsight, that the conveyance really was ambiguous. Where the lessor and a transferee create an ambiguity as to how delay rentals are to be paid, the Oklahoma Supreme Court held in *Superior Oil v. Jackson*, 207 Okl. 437, 250 P.2d 23 (1952), that rental checks could properly name both the lessor and the transferee. As a practical matter, since delay rentals are usually nominal in amount, lessees often pay delay rentals in full to all possible claimants.

The notice of assignment clause can be a double edged sword, as is illustrated by *Atlantic Refining Co. v. Shell Oil Co.*, 217 La. 576, 46 So.2d 907 (1950). There, lessee was not given notice of a transfer of a portion of its lessor's rights. However, the transfer was disclosed in a title opinion and the lessee paid delay rentals based on the title opinion. Unfortunately for the lessee, the title opinion misinterpreted an ambiguity in the conveyance. The Supreme Court of Louisiana held that the lease had terminated because the lessee had no occasion to rely on the public records. If the lease contains provisions protecting the lessee in the event of assignment, the lessee ignores their terms at his peril.

(3) Notice of Nonpayment Clauses

With increasing frequency, leases with "unless" drilling-delay rental clauses contain language to prevent automatic termination in the event of mistake in payment of delay rentals. Notice of nonpayment clauses provide that the lease will not terminate until notice of the incorrect payment or other failure is given and time given for proper performance. They are usually included as part of the paragraph containing the drilling-delay rental clause.

(A) ENFORCEABILITY OF NOTICE OF NONPAYMENT CLAUSES

Unless specifically drafted to apply to delay rental payments, notice of nonpayment clauses will probably be held inapplicable. Even if the notice language refers to payment of delay rentals, a court may refuse to enforce it. In a major Oklahoma case, *Lewis v. Grininger*, 198 Okl. 419, 179 P.2d 463 (1947), the Oklahoma Supreme Court considered a clause that provided:

"It is agreed that neglect or failure to pay rentals when due shall not operate to forfeit or cancel this lease, until lessor gives lessee notice by

> registered mail of said failure to pay rental;
> whereupon lessee shall pay same within 10
> days of receipt of said registered letter, or this
> lease is void."

The Oklahoma Supreme Court refused to give effect to the plain language of the notice clause on
the grounds that it was "in conflict" with the
"unless" drilling-delay rental clause and indefinite in its terms. Similar decisions are found in
Michigan and Montana.

However, the Fifth Circuit Court of Appeals, in
a case arising out of Texas, has upheld a notice of
nonpayment clause. In *Wooley v. Standard Oil
Co.*, 230 F.2d 97 (5th Cir.1957), the following language was added to the delay rental clause:

> "If Lessee shall, in good faith and with reason-
> able diligence, attempt to pay any rental, but
> shall fail to pay or incorrectly pay some portion
> thereof, this lease shall not terminate unless
> Lessee, within thirty (30) days after written no-
> tice of its error or failure, shall fail to rectify
> same."

The court said that it was obvious that the parties
intended to contract against the harsh rule of automatic termination and held that the language
saved the lease. Language similar to that approved in the *Wooley* case is often found in
leases.

(b) Drafting Notice of Nonpayment Clauses

Most commentators agree that the intention of the parties ought to control, though they may differ in their drafting suggestions. Two problems are presented to the draftsman. The first is to draft the language of the lease clearly. Specific provision that the notice terms are to apply to delay rental payments and inclusion of the provision as a part of the delay rental clause ought to be sufficient.

The second problem presented to the draftsman is to avoid the appearance of overreaching. The notice clause in *Lewis v. Grininger*, if applied literally, would have excused the lessee from making any payments of delay rentals until and unless notice was given by the lessor. Such language may well be considered unconscionable. A more limited formulation, such as that of *Wooley*, where the lessee is excused for paying the wrong amount or the wrong person, but not for neglect to pay, is more likely to be upheld.

(4) Use of "Or" Leases

Use of leases with "or" drilling-delay rental provisions has become more popular as leases have become more valuable. Under the terms of an "or" lease, the lessee's rights do not terminate automatically for failure to pay delay rentals properly. The lessee has an obligation throughout the primary term of the lease to drill or pay or surrender the lease. If the lessee fails his obligation, the lessor may sue for the delay rental payment or, under the terms of most leases, institute forfeiture procedures. Generally, forfeiture procedures require notice to the lessee and time for correction of the defaults.

(5) Use of Paid-Up Leases

Another device designed to protect against loss of the lease for failure to pay delay rentals properly is the "paid-up" lease. A paid-up oil and gas lease is one under which all delay rentals bargained for are paid in advance. The lease is held for the full primary term by the initial payment to the lessor.

The choice of a paid-up lease is an economic decision. Factors that increase the likelihood that a paid-up lease will be used are large bonuses, small delay rental payments, short primary terms, and small acreages or fractional interests. The larger the size of the bonus, the more likely a paid up lease will be used because the greater will be the potential loss from a failure to pay delay rentals properly. Small acreages or fractions, short primary terms, and modest delay rentals make the paid-up option more affordable.

Sometimes a paid-up lease is created by striking out the drilling-delay rental clause and noting in an addendum that delay rentals have been prepaid. That is a dangerous procedure. The drilling-delay rental clause of an oil and gas lease generally is drafted so that it "dovetails" with other clauses of the lease. In the Texas lease in the Appendix, for example, the shut-in royalty clause (in paragraph 3), the dry hole clause, and the cessation of production clause (both in paragraph 6) refer to the drilling-delay rental clause to make payment of delay rentals a prerequisite for continuation of the lessee's rights under certain circumstances, or to establish the amount of payments due. If the drilling-delay rental clause is struck from the lease, serious ambiguities may be created.

If a "paid-up" lease is desired, it is preferable to use a lease form drafted for that purpose. Some commonly available forms simply omit the drilling-delay rental clause and conform other provisions of the lease to accommodate the omission. It is better to use a formulation such as the following, from the AAPL approved paid-up lease form for Oklahoma:

> "This is a PAID–UP LEASE. In consideration of the down payment, *Lessor agrees that Lessee shall not be obligated,* except as otherwise provided herein, *to commence or continue any operations during the primary term,* or to make any rental payments during the primary term . . . . " (Emphasis added).

This language makes it clear that the lessee can maintain the oil and gas lease for the primary term without drilling an exploratory well on the premises. Without such language, the courts might resurrect the implied requirement that the oil and gas lessee drill an initial test well on the leased premises within a reasonable period of time.

F. CONCLUSION

The primary goals of the lessee in a modern oil and gas lease are achieved in just three clauses: the granting clause, the term clause, and the drilling-delay rental clause. These clauses are essential to the lease. However, there is much more to most leases, as can be seen from a cursory review of the leases in the Appendix. In Chapter 9 we will consider several kinds of defensive clauses commonly found in leases, and in Chapter 10 we will examine the lease royalty clause.

CHAPTER 9

DEFENSIVE CLAUSES IN OIL AND GAS LEASES

In contrast to real property leases, oil and gas leases are generally interpreted strictly against lessees. Strict interpretation has been justified by the rules of construction that a written instrument is to be interpreted against its drafter, that an instrument is to be construed against the party owing performance under it, and that an option is to be construed against its holder, as well as by public policy in favor of freeing property to be developed by another. At the trial court level, strict interpretation against the lessee may be explained on the mundane basis that disputes over lease provisions usually go to trial in the county where the lessor lives, before judges and juries residing in the area.

Lessees have countered strict interpretation by using defensive language. As a result, modern oil and gas leases contain lengthy and complicated defensive clauses to protect what lessees regard as their legitimate interests. This chapter will examine those clauses and their interpretations.

A. CLAUSES TO EXTEND OR MAINTAIN THE LEASE

Important clauses have been added to oil and gas leases to modify the general rule that a lease terminates at the end of its primary term unless there is production in paying quantities. These clauses extend the term of the lease without production in stated circumstances. This group of defensive clauses includes: (1) dry hole clauses, (2) operations clauses, (3) pooling and unitization clauses, (4) force majeure clauses, (5) shut-in royalty clauses and (6) cessation of production clauses.

1. DRY HOLE CLAUSES

A dry hole clause prevents implication of condemnation or abandonment of the lease from the drilling of an unproductive well on the leased premises and clarifies what the lessee has to do to maintain the lease for the remainder of the primary term. Sometimes the clause gives the lessee a free rental period after drilling a dry hole. It always specifically affirms the lessee's right to maintain the lease for the remainder of the primary term by paying delay rentals. A common formulation is similar to paragraph 6 of the Texas lease in the Appendix:

"If prior to discovery of oil, gas, or other hydrocarbons on this land . . . Lessee should drill a dry hole or holes thereon . . . this Lease shall not terminate if Lessee commences

[*210*]

additional drilling or re-working operations within sixty (60) days thereafter, or if it be within the primary term, commences or resumes the payment or tender of rentals or commences operations for drilling or re-working on or before the rental paying date next ensuing after the expiration of sixty (60) days from the date of completion of the dry hole. . . . "

Dry hole clauses were developed to avoid dispute whether a lessee who drilled a dry hole could maintain the lease for the remainder of the primary term by making delay rental payments. Before oil and gas leases routinely included dry hole clauses, lessors successfully argued on occasion that drilling operations that resulted in a dry hole constituted an irrevocable election of the drilling option of the drilling-delay rental clause. Thus, the only way the lessee could maintain the lease after drilling a dry hole was to continue to drill. Lessees convinced other courts that since the essential consideration for the grant of an oil and gas lease was the conduct of drilling operations, the drilling of a dry hole during the primary term should hold the lease for the remainder of the primary term without either payment of delay rentals or further drilling operations. To avoid confusion, the dry hole clause was developed.

Common problems with dry hole clauses include (a) what is a dry hole, (b) when a dry hole is completed, and (c) when the payment of rentals is due after a dry hole.

a. What Is a Dry Hole

How dry must a well be to qualify as a dry hole? Must it be a "duster," one that locates no oil or gas? Or does the term include a well that is incapable of producing in paying quantities? The courts are divided. Because of the division of authority, many leases define the term.

Professor Eugene Kuntz rationalizes the cases by defining a dry hole as an unsuccessful drilling operation. By this analysis, a well capable of less than paying production will not be a dry hole for purposes of the drilling-delay rental clause; commencement of any well will satisfy the drilling-delay rental clause. But such a well will be a dry hole for purposes of the term clause, at least where production in paying quantities is required.

b. When Is a Dry Hole Completed

Many dry hole clauses, like the one quoted, key the lessee's rights to the time of completion of a dry hole. Unfortunately, "completion" has no certain meaning. However, logic requires that a well should be "completed" as a dry hole when it has been drilled to the depth or formation sought and the lessee determines in good faith exercise of business judgment that the drilling operation is unsuccessful.

c. When Payment Is Due

A recurring problem of dry hole clauses is when payment of delay rentals is due after a dry hole is drilled. Many early dry hole clauses were ambiguous as to whether payments were due at the next lease anniversary date or at the anniversary of the completion of the dry hole. Modern dry hole clauses clearly specify the date for resumption of delay rental payments.

Modern dry hole clauses may present similar difficulties, however. The formulation quoted above illustrates one. It provides that the lessee may continue to maintain the lease after drilling a dry hole by paying delay rentals, but payment is excused on the next delay rental due date if the well is completed as a dry hole within sixty days prior to that date. The rationale for such provisions is that it is difficult for oil company lease administration departments to "gear up" to pay delay rentals with less than a sixty day notice.

Counting the days can be harder than it looks. For example, suppose that the lessee completed a dry hole on the premises on November 4. If the delay rental due date was January 2, is the delay rental payment excused under the dry hole clause quoted? The answer is affirmative, of course. Applying accepted contract interpretation principles, the first partial day (November 4) is not counted. Therefore, there are 26 days through the end of November. December has 31, and both January 1 and January 2 are counted in the total of 59 days. However, if you had any hesitancy at all in reaching this conclusion, the possibility for miscalculation will be apparent.

The interpretation that the parties to the lease give to its provisions is controlling. For example, in *Superior Oil Co. v. Stanolind Oil & Gas Co.*, 230 S.W.2d 346 (Tex.Civ.App.1950), affirmed 150 Tex. 317, 240 S.W.2d 281, the dry hole clause provided:

> "Should the first well drilled on the above described land be a dry hole, then and in that event, if a second well is not commenced on said land within twelve months thereafter, this lease shall terminate as to both parties, unless the lessee on or before the expiration of said twelve months shall resume the payment of rentals in the same amount and in the same manner as hereinbefore provided."

The delay rental anniversary date was March 3. A dry hole was completed on February 3. On the following January 28, the lessee tendered delay rental payment to the lessor "in payment of delay rentals for the period of February 3, 1946 to February 3, 1947"; the lessee interpreted the quoted provision as requiring payment twelve months from the completion of the dry hole. In January, 1947 and 1948, similar payments were made. After making the 1948 payment, the lessee transferred the lease to an assignee. The assignee tendered payment February 5, 1949, assuming that the twelve month period ran from the lease anniversary. The tender was rejected by the lessor, and the lease was held terminated. The reasoning of the intermediate appellate court was that the actions of the lessor and the original lessee had amended the terms of the lease agreement. The supreme court held that the terms were ambiguous so that the interpretation of the original lessor and lessee were controlling. As a result of the *Superior v. Stanolind* decision, assignees of leases commonly require that the administrative files of the assignor (which usually contain cancelled delay rental checks) be delivered with the assignment.

2. OPERATIONS CLAUSES

An operations clause is included in most oil and gas leases to protect the lessee against expiration of the primary term while drilling operations are in progress. As is discussed at pages 173–175, in most states, actual production is required at the end of the primary term to extend the lease. A minority of states, including Kentucky, Oklahoma, and Montana, have held to the contrary on the basis that since the lessee has the right to commence operations at any time during the primary term, he ought to be able to finish what he has begun, if he acts with due diligence.

To avoid dispute over the issue, most modern leases contain a provision specifically extending the lease while operations are in progress. A common formulation from an Oklahoma form follows:

"12. Notwithstanding anything in this lease to the contrary, it is expressly agreed that if the lessee shall commence operations for the drilling of a well at anytime while this lease is in force, this lease shall remain in full force and effect and its term shall continue so long as such operations are prosecuted, and if production results therefrom, then as long as such production continues."

In effect, the operations clause makes drilling operations the equivalent of production for purposes of the term clause. It permits the lessee to extend the lease while completing drilling operations begun before the primary term's expiration.

a.   Well Completion v. Continuous Operations Clauses

The Oklahoma lease quoted permits the lessee to complete drilling operations begun before the end of the primary term, but does not permit him to commence additional operations "if the lessee shall commence operations . . . while this lease is in force . . . its term shall continue so long as *such operations* are prosecuted. . . .") (Emphasis added). The lessee cannot complete a well begun prior to the end of the primary term as a dry hole and then spud another well. His lease would terminate when the drilling operations commenced during the primary term were completed, unless some other language of the lease (such as the dry hole clause) extended the lease. An operations clause that permits only completion of the operations begun before the termination date is often called a *well completion clause.*

[*216*]

A slight modification of the language will create what may be called a *continuous operations clause,* one not restricted to completion of a well in progress at the end of the term:

> "This lease shall continue in force so long as drilling or reworking operations are being continuously prosecuted on said land . . . ; and drilling or reworking operations shall be considered to be continuously prosecuted if not more than sixty (60) days shall elapse between completion or abandonment of one well and the beginning of operations for the drilling or reworking of another well."

This language permits a lessee to commence a well before the end of the primary term, abandon it after the end of the primary term and continue to hold the lease by starting another well within sixty days. The result can also be obtained by modifying the term clause so that it provides that the lease will extend for "so long thereafter as oil or gas are produced . . . or drilling or reworking operations conducted . . . ." The essential difference between continuous operations clauses and well completion clauses is that the former extends the lease so long as any operations take place, whether for the well in progress at the end of the term or another.

Just how crucial the distinction between a well completion clause and a continuous operations clause can be is illustrated by *Sunac Petroleum*

Corp. v. Parkes, 416 S.W.2d 798 (Tex.1967). There, the clause in question provided:

> "5. If prior to discovery of oil or gas on said land Lessee should drill a dry hole or holes thereon . . . this lease shall not terminate if Lessee commences additional drilling or re-working operations within sixty (60) days thereafter . . . If at the expiration of the primary term oil, gas or other mineral is not being produced on said land but Lessee is then engaged in drilling or re-working operations thereon, *the lease shall remain in force so long as operations are prosecuted* with no cessation of more than thirty (30) consecutive days, and if they result in the production of oil, gas or other minerals so long thereafter as oil, gas, or other mineral is produced from said land." (Emphasis added).

Sunac commenced a well shortly before the end of the primary term of the lease on land that had been pooled with the lease for gas only. It was completed after the end of the primary term as an oil well. Recognizing that an oil well on the pooled unit would not satisfy the lease terms, Sunac immediately began a second well, this one located on the leased land. It was completed as a producing oil well. Sunac contended that it was entitled to protection either under the dry hole clause (the first sentence of the quoted paragraph) or the second sentence of the quoted paragraph, which Sunac contended was a continuous operations clause. The Supreme Court of Texas rejected Sunac's argument, holding that the lessee was not entitled to the protection of the dry hole clause because a well that produced oil was not a dry hole, and that what Sunac termed to be a continuous operations clause was only a well completion clause because there was no reference to "additional operations." Therefore, Sunac lost its lease and the producing well.

b. Delay Between Completion of Operations and Production

If the operations clause is read literally, a lease extended beyond the primary term by operations will terminate upon completion of operations unless production follows immediately. Since there is often a substantial delay between completion of operations and actual production, a literal interpretation is contrary to the purpose of the operations clause. Accordingly, some courts have held that an operations clause extends the lease by inference beyond the time that drilling operations are completed for so long as the lessee exercises due diligence in completing, equipping, and producing the well and marketing its production.

Sword v. Rains, 575 F.2d 810 (10th Cir.1978), provides an example of application of the inference. There, Rains commenced a test well on Sword's land in Kansas close to the end of the primary term. The lease was extended beyond the primary term under the terms of its well completion clause. The well was completed and made ready for pipeline connection approximately two weeks later. However, more than eight months passed before gas was sold. Sword sued Rains contending that the lease had expired. The district court and the court of appeals rejected the contention. They held that operations under a well completion clause will extend the lease as long as the lessee diligently pursues putting the lease into actual production.

The conclusion reached by the court in *Sword v. Rains* seems logically unassailable. There will always be a delay between completion of drilling operations and inception of production. The delay may be only a few days for oil production or for production from a gas well located close to a gas pipeline. It may extend for months or years in the case of gas wells not serviced by pipelines or in times of economic recession. If operations clauses in oil and gas leases are to have practical meaning, they should extend the lease so long as diligent efforts to produce and market are being made.

3. POOLING AND UNITIZATION PROVISIONS

"Pooling" and "unitization" are often used synonymously in the oil industry. However, *pooling* is defined as bringing together small tracts or fractional mineral interests for the drilling of a single well for primary production on a spacing unit. In contrast, *unitization* generally refers to combining leases and wells over a producing formation for field-wide operations. Unitization is

almost always associated with pressure maintenance or with secondary or tertiary recovery operations rather than with primary recovery operations. Pooling and unitization clauses in leases
give the lessee authority to commit the lease
property to pooling and unitization and adjust the
rights of the lessor and lessee accordingly.

Without the lessor's approval, the lessee generally may not affect the rights of the lessor under
the lease by pooled or unitized operations. A
lessee who accepts a lease without the pooling
power cannot extend it to its secondary term
without drilling a well on the leased property,
even though spacing rules do not permit drilling
or geological evidence suggests it would be unwise. And, if a lessee pools his interest under a
lease without a pooling clause with that of other
property owners and drills a well on the leased
premises, he must account to his lessor for the
full lease royalty on production from the well,
though his pooling agreement allocates to him only a portion of production from the well. Without
agreement of the lessor, either in the lease or by
separate agreement, the lessee cannot affect the
lessor's rights.

The general rule that pooling or unitization by
the lessee cannot affect the rights of the lessor
under the lease does not fit business realities. It

is frequently a legal or economic necessity that operations combine two or more properties. It is not practicable to seek lessors' approval on a case by case basis because of the administrative costs involved and because of the likelihood of demands for extra compensation. Where compulsory pooling or unitization is not easily achieved, lessees may seek the right to pool or unitize in the lease.

a.　Community Leases

One way for a lessee to obtain the flexibility he desires is to join all of the mineral owners of the relevant property in a single lease. A single lease covering two or more separately owned tracts of land or fractional mineral interests is called a "community lease." In Texas, execution of a community lease pools the lessors' interests as a matter of law, with royalties being apportioned on the basis of the number of acres each lessor has contributed to the community lease. In the other states that have considered the issue, execution of a community lease merely raises an inference that the parties intended to pool their interests.

b.　Pooling Clauses

The most common way that lessees obtain the right to pool their lessor's interests is by a pooling clause in a lease. An example from a Kansas lease follows:

"Lessee at its option, is hereby given the right and power to voluntarily pool or combine the lands covered by this Lease, or any portion thereof, as to oil and gas or either of them, with any other land, lease or leases adjacent thereto, when in Lessee's judgment it is necessary or advisable to do so in order to properly develop and operate said premises, such pooling to be into units not exceeding eighty (80) acres for an oil well plus a tolerance of 10%, and not exceeding six hundred and forty (640) acres for a gas well plus a tolerance of 10%, except that larger units may be created to conform to any spacing or well unit pattern that may be prescribed by governmental authorities having jurisdiction. Lessee shall execute in writing and record in the County records an instrument identifying and describing the pooled acreage. The entire acreage so pooled into units shall be treated for all purposes, except the payment of royalties, as if it were included in this Lease, and *drilling or re-working operations* thereon, *or production of oil and gas* or other hydrocarbons *therefrom* or the completion thereon of a well as a shut-in gas well, *shall be considered for all purposes*, except the payment of royalties, *as if such operations were on or such production were from*, or such completion were on *the lands covered by this Lease*, whether or not the well or wells be located on the premises actually covered by this Lease. In lieu of royalties elsewhere herein specified, including shut-in gas royalties,

> *Lessor shall receive from a unit so formed
> only such portion of the royalty stipulated
> herein as the amount of his acreage placed in
> the unit,* or his royalty interest therein, *bears
> to the total acreage* so pooled." (Emphasis
> added).

A typical pooling clause grants the lessee a
power of attorney to pool the lessor's interests.
It changes the result that would otherwise occur
under the lease in two ways. First, the pooling
clause modifies the term clause of the lease by
providing that operations anywhere on the unit
formed will be considered to be operations on the
leased premises. Second, the pooling clause obligates the lessor to accept royalty proportionate
to the amount of the leased land included in the
pooled unit. Thus, the pooling clause substantially increases the lessee's flexibility.

c.   Unitization Clauses

As is discussed at pages 11–13, drainage of oil
and gas by enhanced recovery operations may not
be protected by the rule of capture. Besides, as
a practical matter, unit operations require cooperation of all owners. Voluntary cooperation may
not be easy to obtain. Unitization for pressure
maintenance or for secondary or tertiary recovery operations is generally beneficial to all owners in the long run, because it may increase the
total amount of production obtained from the
property. However, in the short run, it may be in
the interest of some or all of the owners to refuse

to agree to unitize and to continue to operate with primary recovery methods. In addition, the difficulty of obtaining voluntary agreement is compounded where there is a large number of owners.

While the unitization power may be as important to lessees as the pooling power, unitization clauses in oil and gas leases are relatively uncommon. There are three reasons for their relative rarity. One is that unitization has been economically justified and technologically feasible only in relatively few situations. However, dramatic increases in oil and gas prices have increased the number of those situations and provided an important stimulus to improvement of technology. Another reason is that the focus of lessees when leases are taken is upon primary production. Unitization for secondary or tertiary recovery techniques may be important in the long run, but the long run seems far away when leases are negotiated. Third, the objections that mineral interest owners have to pooling clauses are even more acute with unitization clauses; unitization clauses are not included in leases because the market will not bear them.

Nonetheless, unitization clauses are found in some oil and gas leases. The Colorado lease in the Appendix has one at paragraph 10:

"Lessee may at any time or times unitize all or any part of said land and Lease, or any stratum or strata, with other lands and Leases in the same field so as to constitute a unit or units whenever, in Lessee's judgment, such unitization is required to prevent waste or promote and encourage the conservation of Oil and Gas by any cooperative or unit plan of development or operation; or by a cycling, pressure-maintenance, repressuring or secondary recovery program. Any such unit formed shall comply with the local, State and Federal Laws and with the orders, rules, and regulations of State or Federal regulatory or conservative [sic] agency having jurisdiction. The size of any such unit may be increased by including acreage believed to be productive, and decreased by excluding acreage believed to be unproductive, or where the owners of which do not join the unit, but any such change resulting in an increase or decrease of Lessor's royalty shall not be retroactive. Any such unit may be established, enlarged or diminished and in the absence of production from the unit area, may be abolished and dissolved by filing of record an instrument so declaring, and mailing or tendering to Lessor, or to the Depository Bank, a copy of such instrument. Drilling or re-working operations upon, or production from any part of such units shall be considered for all purposes of this Lease as operations or production from this Lease. Lessee shall allocate to the portion of this Lease included in any such unit a fractional part of production from such unit on any one of the following basis' [sic]: (a) the ratio between the participating acreage in the unit; or, (b) the ratio between the quantity of recoverable production from the land in this Lease in such unit and the total of recoverable production from all such unit [sic]; (c) any basis approved by State or Federal authorities having jurisdiction. Lessor shall be entitled to the royalties in this Lease on the part of the unit production so allocated to that part of this Lease included in such unit and no more."

Like its pooling counterpart, the unitization clause gives the lessee authority to bind the lessor's interests to a unitization plan. It amends the lease term clause by making production or operations anywhere on the unit the equivalent of production or operations on the lease. It amends the lease royalty clause by giving the lessor a royalty calculated on the production from the unit that is allocated to the unit under the unitization agreement rather than on actual production from the leased premises.

d.   Problems Under Pooling and Unitization Clauses

Pooling and unitization clauses frequently give rise to disputes between lessors and lessees. Among the issues are (1) whether the power has been exercised in good faith, (2) whether the power has been exercised at the appropriate time, (3) whether the lessee has a duty to pool, (4) whether exercise of the power cross-conveys property interests, (5) whether non-operating interests have the right to ratify exercise of the power, and (6) whether the power conflicts with the Rule Against Perpetuities.

(1) Exercise in Good Faith

The courts have recognized an implied requirement that the pooling or unitization power be exercised in good faith. Since the purpose of the clauses is to give the lessee flexibility to comply with well spacing requirements and geological realities and to operate efficiently, the power to pool is limited by those purposes. A lessee should not be able, for example, to pool a portion of one leased property with another leased property solely for the purpose of maintaining two leases by the drilling of one well.

The Texas case of *Amoco Production Co. v. Underwood*, 558 S.W.2d 509 (Tex.Civ.App.1977), writ ref. n.r.e., illustrates the point. There, the lessors contended that the lessee had "gerrymandered" a drilling unit of 688 acres which, under the terms of the leases, would extend eight leases covering a total of approximately 2,250 acres. The unit was designated approximately two days prior to the end of the primary terms of several of the leases. It was alleged that some clearly nonproductive property was included in the unit and some clearly productive property was excluded. A jury found that the unit was established in bad faith, and the trial court cancelled the unit and declared that some of the leases had terminated. On appeal, it was held that the question of good faith is an issue of fact, and that the jury had properly decided that the lessee had acted in bad faith on the basis of the configuration of the unit and the timing of the designation.

It should not be assumed that any multi-lease pooling close to the end of the primary term of one or more of the leases is defective. The lessee's duty is one of good faith, not that of a fiduciary. In accordance with general principles of contract interpretation, the pooling or unitization clause should be (and generally is) interpreted broadly. The issue of good faith or bad faith is a question of fact, and the burden of proof is upon the lessor who asserts bad faith.

(2) Time of Exercise

A second common problem encountered under pooling and unitization clauses is whether the unit designation is accomplished on a timely basis. Often, clauses provide specifically that pooling or unitization will be effective when a stipulation to that effect is recorded and notice given to the lessor. Even where the lease terms are not specific, courts have generally held that in order to hold a lease beyond its primary term the lessee must file the pooling designation before the end of the lease primary term. The lessee's intention and good faith are not enough; formal action is required.

The requirement is not unreasonable. A lease pooling or unitization clause gives the lessee a broad scope of discretion. Though the lessee must exercise the pooling power in good faith, bad faith is difficult for a lessor to show. The obligation of formal action is generally not difficult to comply with and may be seen as a procedural safeguard to avoid more indefinite disputes.

(3) Duty to Exercise the Power

Several cases suggest that lessees have a duty to pool in appropriate circumstances. As will be discussed in Chapter 11 in conjunction with covenants implied in leases, lessees are generally held liable to lessors for drainage of leased property only where it is shown that an offset well would be profitable. Some cases have suggested that the lessee may be liable even if the well is not profitable, if he fails to seek to protect his lessor by pooling the lease with the draining property. By this line of reasoning, the breach of the lessee's implied covenant lies in the failure to pool.

The presence in the lease of a clause authorizing the lessee to pool or unitize makes it more likely that the lessee will be found to have a duty to exercise the powers given. The effect of the lease clause is to transfer the lessor's rights with respect to pooling or unitization to the lessee. Having taken those rights from the lessor, the lessee is required to exercise them in good faith and with prudence. Because the lessor no longer has the right to participate in the pooling or unitization process, the lessee may be found to have an obligation to seek pooling or unitization in situations that are less clear cut than would otherwise be the case. Or, courts may look harder at whether the lessee's failure to act was in good faith.

(4) Cross-Conveyance Theory

In Texas, and probably in California, Mississippi, and Illinois, the effect of a community lease or a voluntary pooling, whether by separate agreement or exercise of the lessee's rights under the pooling clause, is to cross-convey interests in the property among the various interest owners. In other words, each person whose interests are affected by the pooling, acquires a proportionate property interest in the land of the others. The theory has been specifically rejected in Oklahoma, and probably in Kansas, Louisiana, Montana, and West Virginia. In these states, pooling is seen as creating contract rights in the various parties affected.

Acceptance of the cross-conveyance theory may have profound effects. Owners of pooled interests become necessary parties to suits involving the land pooled or unitized. Consent of all persons who have either operating or non-operating interests in the property covered is necessary to a pooling or unitization agreement. In addition, cross-conveyancing theory may cause application of the Statute of Frauds, conveyancing statutes, and the rule against perpetuities to unit agreements, and may affect the choice of venue.

[231]

As a result of these problems, it is common in Texas and other states where cross-conveyance theory may apply to include in pooling or unitization agreements or designations express provisions disclaiming cross-conveyancing. Though such devices have a "boot strap" quality about them, they should be given effect where the intention is clear.

(5) The Right of a Non-executive Owner to Ratify

A fifth common problem with pooling or unitization and community leases is whether the owner of a royalty (or some other non-executive interest) in a part of the tract can ratify. Inevitably, the problem arises after a producing well has been completed upon a tract in which the royalty interest owner has no interest. If the royalty owner has the right to ratify, he will become entitled to a share of the production. If he cannot, he will take nothing.

The problem addressed is a variation of the issue of whether the executive right includes the power to pool non-executive interests, discussed at page 96. The Texas courts have held it does not. If the non-executive right is not affected by pooling, it may ratify or reject the executive's action. In several Texas cases, the courts have permitted ratification of a community lease or a pooling agreement by a royalty owner where the royalty owner has acted promptly. *Ruiz v. Martin*, 559 S.W.2d 839 (Tex.Civ.App.1977) error ref.

n.r.e., is an example. There, the nonparticipating royalty interest owner had rights only to a portion of the land covered by the lease. The mineral interest in that portion of the land and the fee interest in the remainder of the property covered by the lease was owned by Martin. Martin granted a single oil and gas lease covering both portions. Subsequently, the lessee completed a gas well on the portion of the leased property in which the royalty owner had no interest. The royalty owner promptly recorded a written ratification of the lease. The court held that the lease by the mineral owner amounted to a proposal to other interest owners in the same property to pool their interests. Ratification of the lease by the owner of a royalty interest in part of the land had the effect of pooling his interest on an acreage basis. He became entitled to share in royalties even though the well was drilled on a portion of land not subject to his royalty interest.

The underlying rationale of such cases is that without the right to ratify the non-executive interest will be subject to manipulation by the executive so that the value of the interest will be lost; e.g., the shape of the unit or the location of the well may be "rigged" to minimize or cut out entirely the royalty interest. The contrary view, adopted in Louisiana, is that the self-interest of the executive and the power of the courts to intervene to protect the non-executive against bad faith or imprudent exercise of the executive right will be sufficient to protect the non-executive.

(6) Conflict With the Rule Against Perpetuities

Occasionally, the argument is advanced that the pooling or unitization clause of a lease is void because it conflicts with the rule against perpetuities. The argument is that the lease is potentially without end, so that the pooling or unitization power may be exercised after the end of all lives in being at its creation. That argument has been rejected by the Fifth Circuit Court of Appeals applying Utah law on the grounds that a pooling clause creates contract rights rather than property rights. It has also been rejected by the Supreme Court of Kansas on the basis that the lease creates a vested present estate that is not subject to the rule. However, the result is not clear, so some pooling or unitization clauses impose a limitation upon exercise of the power specifically conformed to the rule against perpetuities.

e. Pugh Clauses or Freestone Riders

Lessors often resist pooling or unitization clauses in oil and gas leases. From the view of the mineral interest owner, the pooling clause may be seen to give the lessee a dangerous amount of discretion to preserve the lease and affect the amount of royalties. As discussed, typical lease pooling or unitization provisions provide that operations anywhere on the unit established, even though not on the leased premises, will extend the lease to its secondary term. Also they provide for calculation of the lessor's royalty only on the portion of production allocated to the lease, even where the well is located on the lease.

However, oil and gas lessees need the power to pool. Without it, the lease may be lost or economically wasteful actions required to preserve it. At best, the lessee will be subjected to the time and expense of negotiating (and probably paying for) the lessor's approval each time pooling is desired. Further, since most leases in current use contain pooling provisions, leases without pooling provisions are difficult to market.

A compromise often struck between lessors and lessees with respect to the power to pool is what is commonly called a "Pugh" clause or, in Texas, a Freestone rider. A Pugh clause modifies usual pooling language to provide that drilling operations on or production from a pooled unit will not preserve the whole lease. There are many variations. A simple formulation follows:

"Notwithstanding anything to the contrary herein contained, drilling operations on or production from a pooled unit or units established under the provisions of paragraph 4 [the pooling clause] hereof or otherwise embracing land covered hereby and other land shall maintain this lease in force only as to land included in such unit or units. The lease may be maintained in force as to the remainder of the land in any manner herein provided for, provided that if it be by rental payment, rentals shall be payable only on the number of acres not included in such unit or units."

A Pugh clause compromises the objections of the mineral interest owner with the needs of the lessee. It gives the lessee the flexibility to make pooling decisions, but limits the effect of decisions to that portion of the lease included in the unit. The lessor's royalty is proportionate to the amount of his property that is included in the unit. However, unit operations do not affect the remainder of his land. A Pugh clause takes away much of the incentive that lessees might otherwise have to try to hold large tracts of land by creation of small, multi-lease units.

4.   FORCE MAJEURE CLAUSES

a.   In General

Literally, force majeure means superior force. A force majeure clause in an oil and gas lease relieves the lessee of complying with duties imposed by the lease if failure to perform results from one of the causes named. Often, a force majeure clause modifies the term clause by providing that the lease may be extended without production. The Texas lease in the Appendix contains a typical formulation:

"Should Lessee be prevented from complying with any expressed or implied covenant of this Lease, from conducting drilling, or re-working operations thereon or from producing oil and gas or other hydrocarbons therefrom by reason of scarcity of, or inability to obtain or use equipment or material, or by operation of force majeure, or because of any federal or state law or any order, rule or regulation of a governmental authority, then while so prevented, Lessee's obligations to comply with such covenant shall be suspended, and Lessee shall not be liable in damages for failure to comply therewith; and this Lease shall be extended while and so long as lessee is prevented by any such cause from conducting drilling or re-working operations on, or from producing oil and other hydrocarbons from the leased premises; and the time while Lessee is so prevented shall not be counted against the Lessee, anything in this lease to the contrary notwithstanding."

The quoted language excuses failure to perform because of factors beyond the lessee's control. It would maintain the lease if drilling operations were not possible and excuse any breaches of express or implied covenants caused by the factors identified.

b.   Precise Terms Important

The operative factor in force majeure clauses is the breadth of their exculpatory language. Not all clauses are so broad as the one quoted above. A more narrow example is found in paragraph 9 of the Colorado lease in the Appendix:

"Whenever, as a result of any cause reasonably beyond Lessee's control, such as fire, flood, windstorm, or other act of God, decision, law, order, rule, or regulation of any local, State or Federal Government or Governmental Agency, or Court; or inability to secure men, material, or transportation, and Lessee is thereby prevented from complying with any express or implied obligations of this lease, Lessee shall not be liable for damages or forfeiture of this Lease, and Lessee's obligation shall be suspended so long as such cause persists, and Lessee shall have ninety (90) days after the cessation of such cause in which to resume performance of this Lease."

This language would not protect a lessee against loss of his lease for failure to begin drilling operations prior to the end of the primary term. Though its language identifies inability to obtain men and material as one of the implementing factors, it does not expressly refer to extension of the lease without operations or production. The reference to suspension of expressed or implied covenants is not broad enough to protect the lessee because there is no obligation to obtain production on the premises; operations or production is a special limitation to the lease, rather than a covenant.

5. SHUT–IN ROYALTY CLAUSES

The problem of delay between completion and production presented to the court in *Sword v.*

Rains, discussed at pages 219–220 in conjunction with operations clauses, is a recurring problem. It is common for a well to be completed and ready for production but shut in waiting for a market. The majority rule is that a lease terminates at the end of the primary term, even though there is a capability of production; actual production and marketing is required to maintain a lease in a majority of states.

For this reason, most modern oil and gas leases contain a shut-in royalty clause providing for maintenance of the lease by payments in lieu of production if a well capable of producing is shut in. A typical formulation is included in paragraph 3 of the Colorado lease form in the Appendix:

"While there is a gas well or wells on the land covered by this Lease or acreage pooled therewith, whether it be before or after the primary term hereof, and such well or wells are shut in, and there is no other production, drilling operations or other operations being conducted capable of keeping this Lease in force under any of its Provisions, Lessee shall pay as royalty to Lessor (and if it be within the primary term hereof such payment shall be in lieu of delay rentals) the sum of a one dollar ($1) per year per net mineral acre, such payment to be made to the depository bank hereinafter named on or before the anniversary date of this Lease next ensuing after the expiration of ninety (90) days from the date such well or wells are shut-in, and thereafter on the anniversary date of this

Lease during the period such wells are shut-in, and upon such payment it shall be considered that this Lease is maintained in full force and effect."

The effect of the shut-in royalty clause is to make payment of shut-in royalties equivalent to production under the term clause.

a.   Scope of the Clause: Gas or Oil and Gas

Generally shut-in royalty clauses are limited to gas. That is because oil can be temporarily stored at the site of production and then trucked to market. It is not likely that an oil well will be shut in for lack of market. However, there are still some parts of the country in which distances are so great and terrain so rough that oil wells are shut in periodically. Furthermore, conservation commissions sometimes prohibit the flaring of natural gas where there is no available market. This has the effect of "shutting in" oil wells until gas marketing or recycling can be arranged. For these reasons, shut-in royalty clauses should be drafted to apply to both oil and gas, though force majeure provisions will usually protect the lessee.

b. Problems of Interpretation and Administration

(1) What Is a "Shut-in" Well

Can a lessee maintain a lease upon which drilling operations have discovered gas if operations have not been finished? Or, must a well be completed and capable of production in paying quantities before the lessee can avail himself of its protection? The courts have consistently recognized that the clause's major purpose is to substitute payment of the shut-in royalty for actual production where there is no market. Accordingly, capability of production is required.

(2) When Is a Well "Shut-in"

Under some formulations payment of the shut-in royalty is due within a stated period of time after a well is shut-in. When a well becomes "shut-in" then assumes importance. It may also be important where the clause does not specify when payment is due, for then payment in advance of shut-in is usually required. Where the well is actually producing, it is shut-in when it ceases production. If the well has never produced, it is shut-in when completion operations are terminated.

(3) Shut-in for Reasons Other Than Lack of Market

The question frequently arises whether the shut-in royalty clause can be used to maintain a lease from which there is no production for reasons other than lack of a market. For example, assume that a lessee believes that Congress will deregulate natural gas prices. Can he maintain the lease by shut-in payments while he waits to see what happens? Some shut-in royalty clauses are specifically limited to lack of a market. Where market unavailability is not specified, the clause has been held available whatever the cause of the shut-in so long as it is for a good faith business purpose.

(4) How Long May Payments Be Made

The length of time for which shut-in royalty payments may be submitted for production is a common problem. The underlying issue is whether provision for shut-in royalty payments relieves the lessee from the implied obligation to market within a reasonable time. It has generally been held that it does not. This matter is discussed more fully at page 313.

(5) Effect of Failure to Pay

The most troublesome problem with shut-in royalty clauses is the effect of late or inadequate payment. Many cases have turned on classification of the language of the shut-in clause as mandatory or optional. Where the language of the clause is mandatory (lessee *shall* pay . . .), the lessee has breached his promise if payment is not made timely or in the proper amount. The lessor's remedy is to sue for the payment due. The lease does not terminate. In contrast, where the language of the shut-in royalty clause is optional (lessee *may* pay . . .), failure to pay timely or in the proper amount causes termination of the lease because the lessee has failed to meet a condition of the lease.

The mandatory/optional distinction between shut-in royalty clauses can be criticized on at least two grounds. First, however semantically pleasing it may be, this distinction is unlikely to reflect the intention of the parties or what the intention of the parties would have been had they thought about the problem. In the real world, no reasonable business person would opt to relinquish an oil and gas lease capable of production in order to save a nominal amount of shut-in royalties. The first duty of the courts is to give effect to the intention of the parties to the agreement. The intent of the parties to a shut-in royalty clause is not merely to create an option on the part of the lessee, whatever its language may be.

A second objection to the mandatory/optional distinction is that it is often difficult to classify language used. For example, consider the language quoted above. At first glance it looks mandatory because it provides that *"lessee shall pay . . . ."* However, further on in the clause there appears the language "and *upon* such payment it shall be considered that this lease is maintained. . . . " Does the latter quoted phrase change the effect of the language from mandatory to optional? Several courts and Professor Kuntz's treatise suggest that it does.

A more realistic approach to shut-in royalty provisions has been taken by the Oklahoma Supreme Court. In *Gard v. Kaiser*, 582 P.2d 1311 (Okl.1978), that court indicated that it would find termination of a lease for failure to correctly pay shut-in royalties only if the lease clearly indicated that to be the intention of the parties. The reasoning of the Oklahoma court in that case turned upon Oklahoma's rule that capability of production is sufficient to maintain the oil and gas lease. Because marketing of production is not necessary to hold a lease in Oklahoma, it is not essential that a lease include a shut-in royalty clause. Therefore, the Oklahoma court decided it was illogical that the parties intended termination.

6. CESSATION OF PRODUCTION CLAUSES

a. Temporary Cessation of Production Doctrine

It is inevitable that oil and gas wells will cease production from time to time. Equipment must be repaired or replaced periodically. Chemical reactions in the wellbore may require that the well be "reworked." Few oil or gas wells produce constantly over their lives.

A temporary cessation of production will not cause the lease to terminate, despite the literal provisions of the term clause that the lease extends only so long as there is "production." Because of the obvious inequity of termination for a temporary stoppage, the courts have looked to the facts to determine whether cessation was "temporary" or "permanent." Where the lessee moves diligently and promptly to re-establish production or where circumstances excuse inaction, loss of production for substantial periods of time may be termed temporary. For example, in *Saulsberry v. Siegel*, 221 Ark. 152, 252 S.W.2d 834 (1952), a cessation of production for more than four years as the result of a fire at the well was held temporary.

The distinction between temporary cessation and permanent cessation of production is a question of fact. Where the cause of the production stoppage is easily corrected and the lessee is dilatory in taking action, a few months cessation of production may result in loss of the lease.

b. Lease Provisions

Many oil and gas leases contain provisions intended to give lessees more certainty than is given by the temporary cessation of production doctrine. Usually, this takes the form of a temporary cessation of production clause, a provision in the lease that states that the lease will be maintained so long as production does not cease for more than an agreed period of time, usually sixty to ninety days. The Texas lease in the Appendix contains a common formulation in paragraph 6 combined with a dry hole clause:

> "6.   If prior to discovery of oil, gas, or other hydrocarbons on this land, or on acreage pooled therewith . . . production . . . should cease from any cause, this Lease shall not terminate if Lessee commences additional drilling or reworking operations within sixty (60) days thereafter, or if it be within the primary term, commences or resumes payment or tender of rentals or commences operations for drilling or reworking on or before the rental paying date next ensuing after the operation of sixty (60) days from the date of . . . cessation of production."

So long as more than sixty days does not elapse without operations on the property, the lease will not terminate even though there is no production. Such language replaces the "reasonableness" standard of the temporary cessation of production doctrine with the certainty of a definite time.

Like other savings language in oil and gas leases, cessation of production clauses are interpreted strictly against lessees. A recent Oklahoma case suggests that language like that quoted above may contain a trap for unwary lessees where the lease begins to produce in nonpaying quantities. In *Hoyt v. Continental Oil Co.*, 606 P.2d 560 (Okl.1980), the landowner sued for cancellation of a lease that had ceased to produce in paying quantities after expiration of the primary term. Each month for a period of more than a year, operating revenues had totaled less than operating expenses. During that period of time, the lessee had attempted to renegotiate its gas sales contract and studied an expensive completion attempt in a new formation. The plaintiff argued that the cessation of production clause gave the lessee sixty days after cessation of production in paying quantities to act. Continental Oil argued that cessation of production meant a complete cessation of production, not a cessation of production in paying quantities. The Oklahoma Supreme Court found the lease terminated. The court reasoned that during the primary term of the lease, the cessation of production clause modifies the drilling-delay rental clause so that there is no cessation of production unless production ceases entirely. However, after the primary term has expired, the cessation of production clause modifies the term clause and is triggered whenever production ceases to be in paying quantities. Because the lessee failed to resume operations within sixty days after production in paying quantities ceased, the lease terminated.

The implications of this view are profound because of the practicalities of the oil business. Production figures for a particular month are normally not available for fifteen to twenty days after the close of the month. Bills for operating expenses may not be received for even longer periods. By the time production figures and expenses are obtained and operating revenues and costs totalled, the sixty day period may well have run.

In view of their purpose, cessation of production clauses should apply only where there has been a *total* cessation of production. Where operating costs exceed operating revenues for an unreasonable period of time, the lease should terminate by application of the principles (discussed at pages 175–181) governing production in paying quantities rather than by reference to the cessation of production clause.

Of course, the issue can be avoided by use of a cessation of production clause that makes clear that it applies only to total cessation. Paragraph 6 of the Colorado lease in the Appendix is an example:

> "If at any time or times after the Primary Term or before the expiration of the Primary Term all operations, and if producing, *all production* shall cease for any cause, this Lease shall not terminate if Lessee commences or resumes any drilling or re-working operations, or production, within ninety (90) days after such cessation." (Emphasis added).

B. ADMINISTRATIVE CLAUSES

Another group of defensive clauses in lease forms is intended to liberalize the relationship between the lessor and lessee. These are provisions that lessees *could* function without, but their inclusion simplifies administration and helps to avoid problems of strict lease interpretation.

1. PAYMENT OF DELAY RENTALS

Most modern oil and gas leases contain provisions relating to payment of delay rentals that make it easier for lessees to comply with the delay rental clause. Such provisions include nominating a bank as the lessor's agent to accept payment, providing that payment is effective when mailed, and providing that payment may be by check. These provisions are noted in the discussion of the drilling-delay rental clause at pages 198–203.

2. WARRANTY CLAUSES

In several states, a mineral lessor impliedly warrants right to quiet enjoyment of the interest leased unless warranty is expressly excluded or limited. In Texas and some other states, covenants of warranty may be implied from the use of words such as "grant" or "convey" in the granting clause. However, most oil and gas leases contain a specific covenant of title from the lessor to the lessee. A typical lease warranty is found in paragraph 10 of the Texas lease in the Appendix: "Lessor hereby warrants and agrees to defend the title to said lands . . . ."

The language creates only a covenant of warranty, a promise to defend the lessee against future lawful claims and demands. There is no breach until the lessee is physically or constructively ousted from the property. The present covenants of seisin, right to convey, and no encumbrances are not granted because they would expose the lessor to liability for the grant; the lease is ordinarily completed so that the lessor grants 100% of the mineral interest even if he owns only a fraction and existing easements and mortgages are not usually excepted from the warranty. However, some courts have treated the warranty clause of an oil and gas lease as creating full general warranties.

The warranty permits the lessee to recover damages from the lessor if there is a failure of title. In most states, the limit of the lessor's liability will be the lessee's actual damages up to the amount of compensation the lessor has received under the lease plus interest. A warranty clause in a lease also protects the lessee by making available the doctrine of after-acquired title.

The practice of completing lease forms as if the lessor owns 100% of the minerals, discussed at pages 154–155, places a lessor in jeopardy of a breach of warranty; technically, if O grants a lease on Blackacre, in which he owns 50% of the mineral rights, without limiting his grant to the 50% he owns or excepting 50% of the minerals from his warranty, and thereafter his lessee takes a lease from X, the owner of the other 50%, the necessity for the second lease is an ouster of the lessee and a breach of O's warranty. However, the issue is rarely raised because the industry customarily pays bonus only for the fractional interest owned and does not expect to obtain full rights. In view of the industry practice, claim for breach of warranty is not within the intention of the parties. The reasoning of *McMahon v. Christmann*, Supreme Court of Texas, 1957, 157 Tex. 403, 303 S.W.2d 341, discussed at pages 131–132 above, refusing to apply the *Duhig* rule to leases, supports this conclusion.

Warranty provisions of oil and gas leases are frequently struck and disclaimed by lessors. The practice results from increased competition for leases. Until recently, most oil and gas leases were taken after only a cursory review of title. In recent years, rapidly rising lease bonuses have led to careful title searches before leases are taken. Industry response to market conditions make the warranty clause less essential. Moreover, many lessees have felt that suits against lessors for title defects (unless directly caused by the lessors) are an unproductive business practice.

3.   LESSER INTEREST CLAUSE

The lesser interest clause (sometimes called the proportionate reduction clause) has been added to oil and gas leases to protect the lessee against the possibility of being required to pay twice for the same mineral interest. The provision in the

Texas lease form in the Appendix is found within paragraph 10:

> "[I]t is agreed that if Lessor owns an interest in the oil, gas, or other hydrocarbons in or under said land less than the entire fee simple estate, then the royalties and rentals to be paid Lessor shall be reduced proportionately."

The effect of the clause is to permit the lessee to reduce lease benefits to the extent that the lessor owns less than the full mineral interest described. It has also been applied to reduce lease benefits to the lessor by the amount of outstanding non-participating royalty interests.

The warranty clause and the lesser interest clause are mutually supporting. The warranty clause authorizes the lessee to sue the lessor for breach of warranty of title. The lesser interest clause authorizes the lessee to proportionately reduce future lease benefits to the extent that there has been a title failure.

4. SUBROGATION CLAUSE

The subrogation clause empowers the lessee to protect his interest by paying taxes or mortgages encumbering the property and then stepping into the shoes of the former creditors. The subrogation clause is combined with the warranty clause and lesser interest clause in paragraph 10 of the Texas lease in the Appendix:

> "Lessor . . . agrees also that Lessee at its option may discharge any tax, mortgage or oth-

er liens upon said land either in whole or in part, and in the event Lessee does so it shall be subrogated to such lien with the right to enforce same and apply rentals and royalties accruing hereunder towards satisfying same."

The language is overly broad, since it literally would give the lessee the power of subrogation even though liens on the property were not in default. However, oil and gas lessees are not usually interested in acquiring security interests in real property. There is also a possibility that a court would limit the lessee's discretion under the clause by taking into account the purpose of the parties in including it.

5. EQUIPMENT REMOVAL PROVISIONS

Most modern oil and gas leases contain a clause permitting removal of equipment and fixtures after the expiration of the lease. Paragraph 7 of the Texas lease in the Appendix provides that "Lessee shall have the right at any time during or after the expiration of this lease to remove all property and fixtures placed on the premises by Lessee, including the right to draw and remove all casing." The purpose is to give the lessee the broadest possible discretion to determine when to plug and abandon wells and to protect against a finding that equipment left on a lease after its termination has been abandoned.

The courts have generally permitted lessees to recover equipment left on leases after termination even without authorizing lease language.

Moreover, provisions like the one quoted have been given a restrictive interpretation. The weight of authority will not interpret such a clause to permit plugging and abandoning a well capable of commercial production. Likewise, despite the clause language, the lessee must recover his property from the premises within a reasonable time after lease termination.

6.   NOTICE OF ASSIGNMENT CLAUSE

As is discussed at pages 200–203 above, notice of assignment provisions are found in most leases to protect the lessee against the possibility that he will be held to have had constructive notice of assignment by the lessor and thus be required to check the public records before making delay rentals or other payments. The effect of such a clause is to permit the lessee to rely upon the identity of the lessor designated in the lease until he is provided with proof that ownership rights have changed.

7.   NO INCREASE OF BURDEN PROVISIONS

Modern oil and gas leases contain provisions to obviate the possibility that an assignment by a lessor may increase the burden of the lessee's duties under the lease. For example, where the lessor subdivides his property, the lessee may be obligated to provide separate measuring devices and receiving tanks for the production. In para-

graph 8 of the Texas lease in the Appendix is language to deal with this possibility explicitly:

> "No change or division in the ownership of the
> land, rentals or royalties, however accom-
> plished, shall operate to enlarge the obliga-
> tions, or diminish the rights of Lessee
> . . . ."

8. SEPARATE OWNERSHIP CLAUSE

Closely related to no increase of burden provisions is the separate ownership clause. This is to
deal with problems that arise when it is the lessee
who makes an assignment. Where the lessee assigns the lease interest covering separate portions of the tract leased, the failure of the assignee to pay delay rentals on the portion of the lease
assigned will cause the termination of the entire
lease. The whole rental is due on the date
agreed. Paragraph 8 of the Texas lease in the
Appendix contains a typical provision changing
that result, so that the lease will terminate only
as to the part of property for which rental is not
paid:

> "In the event of an assignment hereof in whole
> or in part, liability for breach of any obligation
> issued hereunder shall rest exclusively upon
> the owner of this Lease, or portion thereof,
> who commits such breach."

9. SURRENDER CLAUSE

Another problem that occurs where property
subject to a lease is divided into separate tracts is

when a lease covers a large tract of land and geological or geophysical evidence indicates that only
a portion is potentially valuable. In that circumstance, the lessee may wish to surrender the
lease to the extent that it covers land thought to
be unproductive. General real property - principles will not give him that right; the lease is a
whole and cannot be severed at the lessee's
whim. Modern oil and gas leases modify the general rule by specific provision such as that found
in the last sentence of paragraph 5 of the Texas
lease in the Appendix:

> "Lessee may at any time or times execute and
> deliver to lessor or to depository above named,
> or place of record a release covering any por-
> tion or portions and be relieved of all obliga-
> tions as to the acreage surrendered, and there-
> after the rentals payable hereunder shall be
> reduced in the proportion that the acreage cov-
> ered hereby is reduced by said release or re-
> leases."

10. NOTICE BEFORE FORFEITURE AND JUDICIAL ASCERTAIN-MENT CLAUSES

Many modern oil and gas leases contain clauses
drafted to protect the lessee against lease forfeiture or termination by requiring the lessor to
give the lessee notice of alleged breaches and an
opportunity to correct them. Such provisions are
called notice before forfeiture clauses. Typical

language is found in the first sentence of paragraph 9 of the Texas lease in the Appendix:

"The breach by Lessee of any obligations arising hereunder shall not work a forfeiture or termination of this Lease nor cause a termination or reversion of the estate created hereby nor be grounds for cancellation hereof in whole or in part unless Lessor shall notify Lessee in writing of the facts relied upon in claiming a breach hereof, and Lessee, if in default, shall have sixty (60) days after receipt of such notice in which to commence the compliance with the obligations imposed by virtue of this instrument . . . ."

Judicial ascertainment clauses are closely related to notice before forfeiture clauses, but they go even further to protect lessees. Typically, judicial ascertainment clauses provide that the lease may not be forfeited or declared terminated until the lessor has proved the breach complained of in court and then, after judgment, the lessee has been given a reasonable time to comply. Most lessors' attorneys will reject judicial ascertainment provisions in leases they review.

CHAPTER 10

THE LEASE ROYALTY CLAUSE

The royalty clause is the main provision in an oil and gas lease for compensation for the lessor. The lessor receives a bonus payment for the grant of a lease. During the primary term of a lease, he may receive periodic payments of delay rentals. If production is obtained, he receives royalty, which is usually stated as a percentage of production or the proceeds of its sale free of the costs of production.

Royalty as a percentage of production is a hedge against uncertainty. The existence, as well as the quantity, of oil and gas under a lease is uncertain until drilling. If no production is obtained, the royalty is worthless. If prolific production is found, the royalty will be extremely valuable. A percentage royalty balances the interests of the lessor and lessee against the inherent risks of exploration.

Until recently, the "standard" lease royalty was $1/8$, except in California where it was generally $1/6$. However, the $1/8$ royalty was a victim of the oil boom of the 1970's and 1980's. It is still seen in "wildcat" leasing areas and marginal production areas but $1/6$ or $3/16$ is probably more common. Royalty amounts up to 30% are often negotiated where there is potential for prolific production.

A. COMMON ROYALTY PROVISIONS

Lease royalty usually is a fixed percentage, but it need not be. Sliding scale royalties increasing the percentage payable if production is at high levels are occasionally encountered, particularly in leases from Indian tribes. Provisions for increase of the royalty percentage after the lessee has recovered his costs and for stated minimum royalties are sometimes seen. However, the economic "sense" of all royalty clauses is that the lessor's compensation after production is obtained will vary with the amount of production.

A common formulation is the language of the Texas lease in the Appendix:

"The royalties to be paid by Lessee are as follows: on oil, ⅛th of that produced and saved from said land, the same to be delivered at the wells or to the credit of Lessor into the pipelines to which the wells may be connected. Lessee shall have the option to purchase any royalty oil in its possession, paying the market price therefore prevailing for the field where produced on the date of purchase. On gas, including casinghead gas, condensate or other gaseous substances produced from said land and sold or used off the premises . . . the market value at the well on ⅛th of the gas so sold or used, provided that on gas sold at the wells the royalty shall be ⅛th of the amount realized from such sale."

The provisions for oil royalty assume that the lessor will take his royalty in kind, while the provisions for gas royalty assume that the lessee will dispose of production and then compensate the lessor with a percentage of the proceeds of sale. The difference in structure reflects the physical and economic differences between oil and gas. It is generally possible to store oil economically on the leased premises and sell it on a daily, weekly, or monthly basis. It is uneconomical to store natural gas. Natural gas must be sold into pipelines, which are expensive to construct and maintain. As a result, pipeline companies generally are unwilling to purchase natural gas except on a long term basis; "life of the well" or 15 to 25 year commitments are common. The more gas an operator is able to commit, the better price he is able to obtain (or in a time of surplus of natural gas, the more likely he is to obtain a purchaser). Therefore, gas royalty provisions commonly provide that the lessor will receive his royalty in cash rather than in kind.

B. NATURE OF THE LESSOR'S ROYALTY INTEREST

Except in Louisiana, the lessor's royalty interest under a lease is treated as an interest in real property. After production, both oil and gas are personal property. However, once production takes place, the lessor's rights may be different if he has the right to take production in kind rather than a right to a share of the price for which the production is sold. Where the lessor has a right to a percentage of production as royalty, the royalty right is a reservation from the lease grant, so that the lessor retains title to his share of production. On the other hand, where the lease merely gives the lessor a right to a share of the money from the sale of production, the lessor's interest is a contract right against the lessee. Important economic realities may turn upon the distinction. For example, if the lessee's creditors attach a tank load of oil pursuant to a judgment lien, the attachment may not affect the lessor's royalty share of that oil because the lessor retains title. But attachment of the lessee's checking account before disbursement of gas royalty to the lessor will leave the lessor with merely a contract claim against the lessee.

C. DEDUCTIONS FROM ROYALTY

By definition, royalty is free of costs of production. However, it may be subject to other costs. Sale of natural gas, in particular, may involve substantial costs subsequent to production. There has been much litigation over which of such costs, if any, can be deducted from the lessor's royalty. The trend is toward dealing with the problem by lease provisions.

To understand the problem it is helpful to refer again to the essential differences between oil and natural gas. The lessee generally incurs no large costs after production of oil. Oil can be economically stored on or near the leased premises. It generally requires little cleaning or processing before sale. On the other hand, natural gas often cannot be sold at the wellhead, but must be transported by the lessee to a pipeline. In addition, natural gas may require cleaning, dehydrating, or compressing before sale. Transporting, cleaning, dehydrating and compressing may be very expensive. They may also substantially increase the value of the natural gas.

Some courts have refused to permit deductions from the lessor's gas royalty on the grounds that royalty is free of costs. However, more frequently, some deductions have been permitted. The rationale usually stated for permitting deductions is that the lessor becomes entitled to the royalty when the gas is produced at the wellhead. Therefore, the lessor should share ratably in costs incurred after production. On this basis, the lessee is permitted to "work back" from the price for which the gas is sold to its value at the wellhead to calculate royalty.

Essentially, the distinction made is between *costs of production* and *costs subsequent to production*. The lessee is obligated to pay all costs of production, but the lessor shares proportionately in costs subsequent to production since they are incurred after production and increase the value of production. However, the distinction is often difficult to apply. Generally, all costs incurred on the lease to get oil or gas to the surface and to make it ready for market are treated as costs of production. Thus the cost of separators, gathering lines and storage tanks will be borne by the lessee. In contrast, deductions from royalty are generally permitted for costs of cleaning, dehydration, transportation and production and severance taxes. Courts are divided over compression costs. There is also disagreement over how to treat costs such as depreciation, interest on borrowed money, and overhead.

Of course, the result may be varied by agreement of the parties to the lease. Thus, where the lease royalty clause provides for gas royalties to be calculated on the basis of "proceeds" or "gross proceeds", courts have sometimes been unwilling to permit deductions for costs subsequent to production. Increasingly, lease clauses specifically provide for the lessee's right to deduct certain costs. For example, the following is from a lease used in Texas:

> For gas (including casinghead gas) and all other substances covered hereby, the royalty shall be one-eighth ($^1/_8$) of the proceeds realized by lessee from sale thereof, *less a proportionate part of the costs incurred by lessee in delivering or otherwise making such gas or other substances merchantable* . . . . (Emphasis added).

D. THE MARKET VALUE/PROCEEDS PROBLEM

The market value/proceeds royalty problem has been one of the most widely litigated and expensive problems of the oil and gas industry in recent years. Courts in several states have held that under common royalty clause language lessors are entitled to royalties on natural gas calculated on its market value at the well when delivered, though that may be substantially more than the price for which the gas was actually sold.

1.   THE BASIS OF THE DISPUTE

The market value/proceeds controversy arises from attempts by industry draftsmen to make clear the right to deduct costs subsequent to production from royalty. The problem is best understood by considering the language that has caused the problem:

> "The royalties to be paid by lessee are as follows: . . . *on gas* . . . produced . . . and *sold or used off the premises* . . . the *market value at the well* of one-eighth of the gas so sold or used, provided that *on gas sold at the wells* the royalty shall be one-eighth the *amount realized* from such sale." (Emphasis added.)

The language refers to two methods of calculating gas royalty. It has led to dispute over the meaning of the alternative references.

The intention is that where gas is sold at the wellhead the lessor will receive royalty calculated on the proceeds received by the lessee under the terms of its gas sales contract. However, where gas is not sold at the wellhead, either because there is no market for it there or because there is a better market for it elsewhere, the lessee will have the right to deduct the lessor's proportionate share of additional costs involved to "work back" to the value of the gas at the wellhead. The expectation was that "market value at the well" of natural gas would be the amount realized from the sale of the gas less the lessor's share of costs of transportation, dehydration, compression, and cleaning and processing.

2. DIVISION OF THE CASES

The Supreme Courts of Texas and Kansas have interpreted such language with a twist that was not expected by industry draftsmen. They have held that references to "market value at the well" impose an obligation on the lessee to pay royalties based upon the market value of natural gas *when it is delivered*, rather than upon the amount realized less costs subsequent to production. Since the economics of the gas industry require that gas be sold under long term contracts, much of the gas sold on any given day is sold under old contracts at prices that may be substantially less than current market value. Therefore, the decisions have exposed the industry to potential liabilities of hundreds of millions of dollars.

The decision of the Texas Supreme Court in *Exxon v. Middleton*, 613 S.W.2d 240 (Tex.1981) is a good example of the reasoning of the Texas and Kansas courts. In that case, Exxon and Sun Oil Company gathered natural gas from leases in the Anahuac field near Houston, processed the natural gas at a cleaning plant on one of the leases in the field, and then sold the gas "at the tailgate" of the plant under a long term contract. The contract price for the gas was approximately 52¢ per thousand cubic feet. The market price for new deliveries of gas commenced in the area was in excess of $2.00 per thousand cubic feet. The royalty clause of the lessors' lease was similar to the one quoted above. The lessors contended that because the natural gas was sold off the leased premises they were entitled to be paid royalties on the higher price. Reasoning that the plain language must be given effect and relying upon earlier decisions, the Texas Supreme Court sustained the position of the landowners. The Supreme Court of Kansas in *Lightcap v. Mobil Oil Corp.*, 221 Kan. 448, 562 P.2d 1 (1977), *cert. denied*, 434 U.S. 976, followed similar reasoning. *Montana Power Co. v. Kravik*, 189 Mont. 87, 586 P.2d 298 (1978) and *West v. Alpar Resources Inc.*, 298 N.W.2d 484 (N.D.1980), and *Piney Woods Country Life School* v. *Shell Oil Co.*, 539 F.Supp. 957 (D.C.Miss.1982), suggest that the Supreme Courts of Montana, North Dakota and Mississippi would reach the same result.

In contrast, the Supreme Court of Oklahoma in *Tara Petroleum Corp. v. Hughey*, 630 P.2d 1269 (Okl.1982), the Louisiana Supreme Court in *Henry v. Ballard & Cordell Corp.*, 418 So.2d 1334 (La.1982), and the Arkansas Supreme Court in *Hillard v. Stephens*, 276 Ark. 545, 637 S.W.2d 581 (1982), have rejected the argument that lessees should pay royalties for natural gas calculated upon values greater than the amounts received. The reasoning of these courts is that since the lessee has an implied covenant to market gas within a reasonable time and at the best available price, the intention of the lessor and the lessee under the oil and gas lease is that gas royalty should be calculated on the price received. Essentially, these courts have recognized the purpose of the alternative language in gas royalty clauses to charge back to lessors their share of costs subsequent to production.

The issue remains to be determined in many states. In addition, there are important subsidiary issues that have not yet been resolved in the states where the market value at the well of natural gas sold can be greater than the price for which it is sold. Among these are how market value is to be determined and what kinds of curative or avoidance devices can be used.

3. HOW TO DETERMINE MARKET VALUE

Since the market value of gas at the well is not limited by the amount realized from sale in Texas and Kansas, market value cannot be determined by "working back" from the contract price. Therefore, the courts have had to address the issue of how to determine market value at the well. They have generally done so by reference to "comparable sales." In *Texas Oil and Gas Corp. v. Vela*, 429 S.W.2d 866 (Tex.1968), the first of the market value royalty cases, the Texas Supreme Court said that the market value of gas was to be established by reference to comparable sales, "sales of gas comparable in time, quality, and availability to marketing outlets."

The standard has proven difficult to apply. There has been dispute over what period of time and from what area contracts should be considered as comparable. The issue has been treated as a question of fact within the discretion of the trial judge. Thus, in *Exxon v. Middleton,* the court considered evidence drawn from more than 30,000 gas contracts covering natural gas sold over a substantial portion of the Texas Gulf Coast. It accepted the testimony of experts that the quarterly average of the three highest prices for gas sold in that area was the market value of the gas at the well.

Another problem has been whether the market value at the well can exceed the price ceiling established by government regulations. For example, can the market value at the well of gas produced from a well certified under § 103 of the Natural Gas Policy Act of 1978 exceed the applicable § 103 price ceiling? Here, the courts of Texas and Kansas have reached different results. In *Lightcap v. Mobil Oil Corp.,* the Kansas Supreme Court held that it could. It reasoned that "market value at the well" is a reference to a theoretical free market price. If lessees contractually obligate themselves to pay gas royalties on a theoretical free market price, the courts cannot protect them against their folly merely because of legislated price ceilings. On the other hand, the Texas Supreme Court in *First National Bank in Weatherford v. Exxon Corp.*, 622 S.W.2d 80 (Tex.1981), held that comparable sales must be of similar legal quality, so that the market value at the well could not be greater than the maximum government regulated price for gas from that well.

4.   CURING OR AVOIDING THE MARKET VALUE/PROCEEDS ROYALTY PROBLEM

A number of devices have been used to cure or to avoid the market value royalty problem. These have included (a) seeking relief in the form of higher prices for the natural gas sold, (b) amending existing leases or selecting lease forms for current use with royalty language that does not refer to "market value", and (c) use of division orders.

a.   Seeking a Compensating Higher Price

Some lessees, squeezed between the obligation to pay royalty for gas on its current market value and a commitment to sell the gas at a low price, have sought relief in higher prices. Old gas contracts have been amended and new ones drafted to include "excess royalty" clauses that solve the market value/proceeds royalty problem by increasing the sales price to the extent necessary to offset the higher royalty. There are at least three problems with this approach. First, it is necessary to persuade a gas purchaser to accept an "excess royalty" clause in its gas sales contract. In times of gas surplus, that may be difficult to do. Second, if the gas purchaser agrees to an excess royalty clause, it will probably condition its obligations under the clause on agreement of appropriate regulatory commissions to include the excess royalty in the pipeline's cost of service, which is passed onto the ultimate consumer. It is not certain that the Federal Energy Regulatory Commission and state regulatory bodies will permit pass through as a matter of course. Third, any excess royalty clause will be subject to the limits of the Natural Gas Policy Act of 1978; the total price received by the lessee must not exceed NGPA limits.

For those producers with leases with market value royalty clauses who have committed natural gas to interstate commerce under the Natural Gas Act of 1938, there is the possibility of "special" rate relief under § 7 of the Natural Gas Act. That possibility may be illusory, however. In *FERC v. Pennzoil Producing Co.*, 439 U.S. 508 (1979), the U.S. Supreme Court held that FERC may grant special rate relief, but is not required to do so unless the impact of the excess royalty is so great that it makes the producer's operations unprofitable. By that time, of course, the typical oil and gas lease will have terminated for lack of "production in paying quantities." It is as yet unclear how strict a position FERC will take.

b. Specific Lease Language

The market value royalty problem can also be cured or avoided by provisions in the lease. The easiest way is to use a gas royalty clause couched in terms of "proceeds" or "amount realized." For example:

> "To pay lessor for gas . . . produced and sold or used off the leased premises . . . one-eighth ($^1/_8$) of the *gross proceeds* received . . . ." (Emphasis added).

The major problem with the use of a proceeds royalty clause is that it may preclude the lessee from deducting the lessor's share of costs subsequent to production from royalty payments. There is precedent that provision for royalties on "proceeds" or "gross proceeds" means that no deductions for transportation, compression, cleaning or processing may be made unless they are specifically permitted in the lease language or the lease specifies that proceeds are to be calculated at the well.

Another approach occasionally seen is to make the gas royalty payable to the lessor in kind, as is usually done with oil royalty. Of course, it is impractical for the lessor to take the gas royalty in kind because of the physical nature of gas. Such a clause forces the lessor to make his own arrangements for sale, which prevents him from complaining about the arrangements made by the lessee. However, the approach sacrifices the leverage the lessee is given in negotiating favorable sales contracts by having the lessor's gas available for sale. It may also lead to gas balancing problems, which are discussed at pages 361–363, if the lessor does not sell his gas or if the lessor's purchaser does not take the lessor's proportionate share of the gas.

A more satisfactory formulation for the lessee is one that couches the obligation to pay royalties in terms of proceeds and that spells out the lessee's right to deduct the lessor's share of costs subsequent to production. The royalty clause quoted in the discussion of deductions from royalties, at page 263 above, is a good example. Such a clause restates what "market value at the well" was intended to mean. If the sale takes place at the well, the lessor will be paid royalty on the amount realized by the lessee under its gas contract. If the lessee negotiates a sale off the premises, he can charge the royalty interest with costs subsequent to production and work back to the market value at the well.

c.   Use of Division Orders

A third device for curing the market value problem is amendment of the oil and gas lease or ratification of the gas contract by division order provisions. A gas division order is an agreement by those entitled to share in production proceeds as to how the funds should be distributed. Division orders are used in the gas industry to protect the purchaser of production or the lessee who distributes proceeds of sale against being "caught in the middle." They bar the co-owners from suing if they subsequently disagree with the proportions in which the proceeds were divided.

Some lessees have attempted to solve market value royalty problems by division order provisions that bind lessors to accept the amount realized by the lessee in settlement of royalty claims. For example, in *Exxon v. Middleton,* the lessors had signed a division order that provided:

> "The undersigned agree to accept payments so made in full payment of the royalties due them . . . ."

>   . . . .

> "2. The royalties payable to the undersigned on gas produced and saved from said lease or unit shall be computed on the value of the quantities marketed . . . such value to be determined as follows:

>   . . . .

> "B. The value of gas processed . . . shall be the sum of (1) the proceeds derived from the sale of such liquids . . . plus (2) the proceeds derived from the sale of residue gas . . . ."

If effective, such language transforms the market value royalty clause into a proceeds royalty clause. In *Exxon v. Middleton,* the Texas Supreme Court held that the quoted language bound the lessors until they revoked their division orders. Apparently, the court's decision was based on a theory of estoppel rather than contract, for it held that the division order could be revoked even though there was language that suggested they were irrevocable.

The inherent problem of using a division order to try to cure the market value royalty problem is whether the lessor understands the "fine print" of the agreement. If he does not, it may be set aside. If he does, it is difficult to comprehend why he would sign it.

E. FAILURE TO PAY ROYALTY

In general, a lessor's remedy against a lessee who fails to pay royalty is to sue for the royalty plus interest at the statutory rate. The lease will not terminate or be cancelled for nonpayment of royalty. There are two reasons for the reticence of courts to cancel leases for royalty nonpayment. One is the structure of the royalty clause. In contrast to the drilling-delay rental clause, the royalty clause is structured as a covenant from the lessee to the lessor, and the usual remedy for a breach of promise is a suit for damages. The second reason is that damages equal to the royalty due plus interest are adequate to make the lessor whole.

The general principle that the courts will not terminate leases for nonpayment of royalties is limited by the inherent power of the courts to do equity. The courts do not lack the power to terminate leases for nonpayment of royalties, they merely refrain from exercising it. Where, for example, a lessee knowingly withheld lessor's royalties for speculative purposes, a court might properly decide to exercise its power.

The power of the courts to cancel leases for nonpayment of royalties is explicit by statute in North Dakota and Louisiana. In North Dakota, a statute specifically authorizes a court to determine "that the equities of the case require cancellation" if royalties are not paid or improperly paid. In Louisiana, prior to the 1975 Mineral Code, cancellation of leases for nonpayment of royalties was common where an appreciable period of time had passed without payment of royalties and no justification for the delay was shown. The Louisiana Mineral Code (Articles 137–142) limits the availability of cancellation to situations where the lessor has been defrauded or where the court finds cancellation necessary to do equity. It also seeks to provide a meaningful remedy for the lessor short of cancellation, prescribing damages of double the amount due plus attorneys fees.

CHAPTER 11

IMPLIED COVENANTS IN OIL AND GAS LEASES

Modern oil and gas leases are drafted and prepared by lessees to protect the interests of the oil industry. Ordinarily, however, lessees are held bound by implied terms in addition to those that are written. Implied covenants in oil and gas leases are unwritten promises that generally impose burdens on lessees and protect lessors. This chapter will examine the basis and application of common implied covenants.

A. THE BASIS OF IMPLIED COVENANTS

1. IMPLIED IN FACT OR IN LAW?

There is substantial debate whether implied covenants are implied in fact or in law. Those arguing that covenants are implied in fact do so on the basis that the lease does not state the entire agreement of the parties. As discussed in Chapter 8, the primary goals of the lease are to give the lessee the right to hold the lease during the primary term without development and to permit the lease to be maintained after production for as long as it is profitable. There is little in a typical lease dealing with any other issue. Lease forms do not usually set standards for operation of the property and marketing after initial development. On that basis, some have concluded that where implied covenants are recognized, they reflect the unexpressed intention of the parties; that they are implied in fact.

The alternative view is that implied covenants are implied at law to correct an imbalance of bargaining power. Though the parties to the lease may not have agreed specifically upon the terms of the implied covenants—indeed they may not even have considered the potential issues—implied covenants impose duties upon lessees to achieve a fair, equitable and just result. By this view, implied covenants are legal fictions imposed by law.

A synthesis of these theories has been suggested by Professors Howard Williams and Charles Meyers. They argue that covenants are implied both in fact and law from the contract law principle of cooperation, which requires that parties to a contract cooperate to effect its purposes. Since an oil and gas lease is a contract as well as a conveyance, the principle of cooperation requires certain conduct of the parties both as a matter of public policy (implied in law) and because the conduct was probably intended by the parties when the contract was formed (implied in fact).

2. SIGNIFICANCE OF THE DISTINCTION

The debate over the basis of implied lease covenants is so interesting that it is easy to forget why it is important. The conclusion as to the basis of implied covenants has little effect upon the substantive rights implied, but it may have bearing upon procedural aspects of a dispute. What statute of limitations is applicable, whether the original lessee is liable for breaches that occur after he has assigned the lease, and where an action should be filed for breach (the venue) may depend upon whether covenants are implied in fact or in law.

Perhaps most important, if covenants are implied in fact rather than in law they may be more easily disclaimed by agreement between the parties. Promises implied in fact concern issues that were within contemplation of the parties but not directly addressed in their agreement. If the parties address all of the issues in their agreement, then it would be improper to imply covenants, for they would duplicate or conflict with express covenants. Likewise, if the parties have specifically agreed that there will be no implied covenants or that certain covenants will not be implied, their agreement should stand if they understood the terms of the disclaimers. On the other hand, if covenants are implied at law to achieve a fair, equitable and just result, specific disclaimers of implied covenants need not be given effect. To do so would conflict with the rationale for recognizing the covenants in the first place.

The courts have not often addressed directly the implied in fact/implied in law issue. Implicit in most decisions on implied covenants is the premise that they are implied in fact to "fill in" the agreement of the parties. However, that assumption may be embraced without good reason or because it yields a pleasing result.

B. THE REASONABLE PRUDENT OPERATOR STANDARD

Underlying all implied covenants is the reasonable prudent operator standard. This standard requires the lessee to conduct himself as would a reasonable prudent operator under the circumstances. The standard arises from the nature of the leasing transaction. A grant of an oil and gas lease is an economic transaction entered into by the parties in the expectation of profit. It follows that the lessee should be required to exercise the authority given by the lease as would a reasonable and prudent operator.

The reasonable prudent operator standard was originally formulated to make it clear that the lessee's obligation is less than that of a fiduciary but more than an obligation to act in good faith. However, the standard has developed a life of its own. What it means in a particular circumstance is a question of fact, but it has generally been employed to impose upon a lessee the obligation to act (1) as a competent oilman and (2) with due regard for the lessor's interests, as well as in good faith.

Some have suggested that there is in fact only one implied covenant—the implied promise of the lessee to act as a reasonable prudent operator. A unified analysis does not fit the case method by which the courts have developed the various implied covenants. However, it is a useful approach, because the reasonable prudent operator standard is the common denominator of all implied covenants.

1. COMPETENCE

The reasonable prudent operator is reasonable and prudent in matters relating to the technology and operating practices of the oil and gas industry. By exercising the operating rights given by the lease, he represents himself to possess an expertise that most persons do not have. Failure to operate competently may bring liability.

The requirement that the operator be competent presents problems akin to those of professional malpractice. Although the operator is not an insurer of operating success, he is required to conduct himself as would other members of the industry in similar circumstances. But is that standard a local standard, or a regional standard, or a national standard? The issue is an important one because the level of technological sophistication of oil companies varies substantially depending on their size and the region in which they operate. The trend is toward treating the standard of competence like the professional liability standard, which imposes a regional or even national standard rather than a local one.

2. WITH DUE REGARD FOR THE LESSOR'S INTERESTS

The reasonable prudent operator must consider his lessor's interests while pursuing his own. He does not owe a fiduciary duty to the lessor, and liability does not necessarily follow from a bad decision. Indeed, the lessee's decisions may have a foreseeable adverse impact upon the lessor without triggering liability. However, the duty requires more than good faith. What is required is that the lessee make decisions with due regard to the interests of the lessor and to the nature of the long term business relationship between the lessee and the lessor. The requirement is based upon business realities. A prudent business person will take into account the interests of those associated with him in transactions, because to do so is necessary to ensure business with them and others in the future.

The determination is objective and often hinges on negative inferences. If it appears that the lessee acted for the purpose of harming or taking advantage of the lessor, he will be held in breach. If it appears that he has speculated with the lessor's interests by using the rights granted by the lease to accomplish some goal not connected with its business purpose, he will be held liable.

C. COMMON IMPLIED COVENANTS

There are at least six commonly encountered implied covenants, but the list that follows is not all inclusive. Whether covenants are implied in fact or in law, and whether they exist independently of one another or merely as applications of the reasonable prudent operator standard, the courts may articulate new obligations. The six implied covenants commonly encountered are: (1) the covenant to test, (2) the covenant to reasonably develop, (3) the covenant to further explore, (4) the covenant to protect against drainage, (5) the covenant to market and (6) the covenant to operate with reasonable care and due diligence.

1. THE IMPLIED COVENANT TO TEST

The implied covenant to test arose in the last quarter of the nineteenth century when fixed term leases were common. The covenant required lessees to test the premises within a reasonable time after the grant of the lease. It was justified as a part of the consideration for the lease.

As discussed at pages 181–182, modern day oil and gas leases avoid the implied covenant to test by providing specifically that the lessee may hold the lease for the primary term without drilling by paying delay rentals. The courts have generally held that, if the lease contains a drilling-delay rental clause, it would be inconsistent to imply a covenant to test.

2. THE IMPLIED COVENANT TO REASONABLY DEVELOP

The implied covenant to reasonably develop is the corollary to the covenant to test. Though no covenant to test arises under modern leases, once a lessee discovers oil or gas, the courts recognize an obligation to continue to develop reasonably.

What is reasonable development is a question of fact that depends on the particular circumstances presented. The essential concept is that the economically motivated prudent operator will fully develop resources under his control within a reasonable time. Failure to do so deprives the lessor of the use of the royalty that he otherwise would have received and suggests that the lessee lacks the economic motivation of the prudent business person.

a. Elements of Proof of Breach

Since the lease gives the lessee the right to make decisions relating to the property, the burden of proof is upon the lessor to show (1) that additional development probably would have been economically viable and (2) that the lessee has acted imprudently in failing to develop. However, in Oklahoma and perhaps a few other states, the burden of proof will shift to the lessee where an unreasonable period of time elapses following initial discovery.

(1) Probability of Profit

In order to prove that additional development should have taken place it must be shown that the lessee would probably have been able to recover his drilling and operating costs plus a reasonable return on investment. The reasonable prudent operator will not drill a well without expecting to make a profit.

Note that it is not necessary to show that profitability is more likely than not. The magnitude of the potential profit must be considered along with the probability. For example, though a particular well may have only one chance in four of success, a reasonable prudent operator would willingly proceed to drill if he could expect a twenty to one return in the first year in the event of success.

Geological testimony can be used to establish probable sites for an additional well. Technological information from nearby wells can be used to establish a probability of productive capacity. Financial statistics of production prices will bear on the profit potential. A variety of evidence may be relevant.

(2) Imprudent Operator

Establishing the second element—that the lessee has acted as an imprudent operator in failing to undertake additional development—is more difficult because it calls into question the lessee's judgment concerning timing of development. Frequently, an operator accused of breaching the implied covenant counters by contending that the profitability of an additional well can be maximized by deferring development for another few years. Judges and juries are not quick to set aside the judgment of businessmen in technologically complicated and high risk ventures.

Each case turns on the facts found. Often, liability seems based upon an inference that the lessee was incompetent or motivated by speculative purposes. Successful development of nearby properties may suggest breach. A similar inference may arise from an extended period of time without development. Some courts have given weight to the willingness of other operators to drill or to the lessee's attitude toward further development, as well. Such decisions usually seem fair on the basis of their facts, but offer few guiding principles. The conclusion that the covenant has been breached often seems to be an exercise of gastronomic jurisprudence.

A case in point is *Waseco Chemical & Supply Co. v. Bayou State Oil Corp.*, 371 So.2d 305 (La. App.), *writ denied* 374 So.2d 656 (1979). The defendant in *Waseco* had leased approximately 80 acres in 1952. At that time, there were fifty wells on the tract, most of which were producing. Recovery averaged more than forty barrels per day of a heavy, asphaltic, high viscosity oil. The defendant made no capital expenditures on the lease and drilled no new wells, conducted no tech-

nical studies, and made no plans for additional development. By 1976, only nine wells were operating and average production from the property
had declined to approximately six barrels per day.
In cancelling the lease, the trial judge noted that
over the same period of time other lessees in the
same field had drilled hundreds of wells and had
substantially increased their production by use of
fire-flooding techniques. The court of appeal upheld the decision, referring to a plat map of the
area showing the development on adjoining lands
and quoting the trial judge who had suggested
that the defendant's management had learned as
much about enhanced recovery operations in the
two weeks of trial as in its twenty years as operator. An inference of incompetence or bad faith
speculation was the basis for finding breach of
the implied covenant.

b. Stumbling Blocks to Enforcement

(1) Notice to the Lessee

In addition to proving the two elements of
breach, the lessor faces additional requirements
to enforce the covenant for reasonable development. One requirement in most states as a prerequisite to requesting lease cancellation is that
demand for development be given by the complaining lessor to the operator. A second requirement is that the lessee be allowed reasonable time following the notice to take action. Both
requirements are based on the principle of coop-

eration. Only when the operator knew or ought
to have known that additional development was
necessary may they be waived. In some states,
including Louisiana, a formal notice is generally
required even to claim damages.

(2) Disclaimer or Limitation in the Lease

Another stumbling block to enforcement may
be lease provisions restricting the scope of the
implied covenant. For example, the Texas lease
form in the Appendix provides in paragraph 9:

> "After the discovery of oil, gas, or other hydro-
> carbons in paying quantities on the lands cov-
> ered by this lease, or pooled therewith Lessee
> shall reasonably develop the acreage retained
> hereunder, but in discharging this obligation
> Lessee shall not be required to drill more than
> one well per eighty (80) acres of area retained
> hereunder and capable of producing oil in pay-
> ing quantities, and one well per six hundred
> forty (640) acres of the area retained hereunder
> and capable of producing gas or other hydro-
> carbons in paying quantities, plus a tolerance
> of ten percent in the case of either an oil well
> or a gas well."

This language attempts to limit the scope of the
implied covenant for reasonable development.
Oil wells are often developed on less than 80 acre
tracts and gas wells on less than 640 acre tracts,
but many courts would be unwilling to find an implied promise by the lessee to drill on denser

spacing when the lease agreement specifically provides otherwise.

Lease terms may also limit the scope of the implied covenant for reasonable development by implication. A good example of an implied limitation is *Gulf Production Co. v. Kishi*, 129 Tex. 487, 103 S.W.2d 965 (1937). Kishi asserted that Gulf had breached its implied covenant for reasonable development on leases of 150 acres and 20 acres. Both leases had addenda that stipulated the number of wells to be drilled following a successful discovery well. The addenda called for 12 wells on the 150 acre tract and 4 wells on the 20 acre tract. In fact, Gulf drilled 15 wells on the 150 acre tract and 6 wells on the 20 acre tract, and all but 3 of the 21 wells produced large quantities of oil. Kishi contended that the implied covenant for reasonable development required development beyond the 16 wells stipulated in the leases, and the trial court agreed. However, in a decision upheld by the Texas Supreme Court, the court of civil appeals reversed the trial court on the grounds that the implied covenant arises "out of necessity and in the absence of an express stipulation" for development. The courts reasoned that since Kishi's leases provided for development, no implied covenant for reasonable development arose. Thus, a lessee's commitment to a development schedule is likely to obviate the implied covenant for reasonable development unless it is clear that the parties intend that the drilling schedule states only a minimum obligation.

An example of a lease with a specific drilling obligation that did not limit the implied covenant is found in *Sinclair Oil & Gas Co. v. Masterson,* 271 F.2d 310 (5th Cir.1959), *cert. denied* 362 U.S. 952 (1960). In *Sinclair,* three leases covering more than 40,000 acres contained express agreements to drill six wells. The lessee asserted that because of the express drilling schedule, there could be no implied obligation to drill more. The court rejected that argument on the grounds that it was not reasonable to conclude that the parties intended that a total of six wells on 40,000 acres would constitute full development. The drilling schedule was interpreted as a minimum requirement rather than an agreed definition of reasonable development.

c.   Remedies for Breach

In general, courts apply one or more of three remedies for breach of the covenant for reasonable development: (1) lease cancellation; (2) conditional lease cancellation; or (3) damages. Favored remedies vary among the states.

(1) Cancellation

Though damages are the preferred remedy for breached agreements, they are not generally the sole remedy for breaches of implied covenants in oil and gas leases. In some states, including Louisiana, lease cancellation is the usual remedy. One rationale for not limiting remedy for breach

of the covenant for reasonable development to
damages is that the covenant is a condition of the
lease; development is the reason for the lease.
Therefore, cancellation is a more appropriate
remedy. More often, cancellation has been justified on the basis of equitable principles; damages
are held to be an inadequate remedy where the
lessee has been found to be incompetent or guilty
of speculation.

Where the breach of the covenant is blatant,
the courts may order outright cancellation of the
lease. Usually, the order is for partial cancellation that excepts portions of the lease surrounding producing wells. However, a cancellation order may cover the whole lease, including
producing wells. For example, the lessee in
*Waseco Chemical & Supply Co. v. Bayou State
Oil Corp.*, discussed above, lost its lease outright
even though it had several producing wells.

(2) Conditional Cancellation

A decree of conditional cancellation is a more
moderate remedy than outright cancellation, and
one which is viewed with growing favor. Under
a decree for conditional cancellation, the lessee is
ordered to commence additional development
within a stated period or suffer cancellation of
the lease.

(3) Damages

Damages may be awarded in addition to a decree of cancellation or conditional cancellation.

Where damages are awarded they are generally measured by lost royalties, those that would have been paid had a well been drilled. If the evidence shows that a well should have been drilled and that it would probably have produced 100 barrels of oil daily, and the lessor's royalty share was one-eighth, damages would be the value of twelve and one-half barrels per day for the period of the breach.

Professors Howard Williams and Charles Meyers have argued convincingly that unless there has been drainage from the premises, the appropriate remedy for the lessor should be interest on the foregone royalties rather than the royalties themselves. Otherwise, the lessor receives a double recovery: damages in the amount of royalties that he should have received plus the royalty itself when the premises are developed and production takes place. Recognizing the double recovery problem, the court in *Cotiga Development Co. v. United Fuel Gas Co.*, 147 W.Va. 484, 128 S.E.2d 626 (1962), permitted the lessor to recover damages based on the foregone royalties, but provided in its decree that future royalty payments be offset by the damages when and if the lessee secured production.

3. THE IMPLIED COVENANT FOR FURTHER EXPLORATION

The implied covenant for further exploration is like the implied covenant for reasonable development in that it imposes obligations on the lessee only after initial development has taken place on the lease. Since the drilling-delay rental clause gives the lessee the right to hold the lease during its primary term without development, there is no implied duty either to reasonably develop or to explore further while the lease is held by delay rentals.

a. Distinguished From the Covenant for Reasonable Development

The implied covenant for further exploration is different from the covenant for reasonable development in the nature of the lessee's complaint. Where the lessor complains of breach of the implied covenant for further exploration, he argues that the lessee has not explored undeveloped parts or formations, rather than that the lessee has failed to develop known deposits.

There has been considerable debate whether the law recognizes an implied covenant for further exploration separate from the covenant for reasonable development. Charles Meyers first identified the covenant for further exploration as a separate implied promise. His analysis has been challenged by a number of writers, including Earl A. Brown, author of a respected treatise on oil and gas leases. Since then, his analysis has been rejected specifically by the Supreme Court of Oklahoma and implicitly by the Supreme Court of Texas.

The major focus of the debate is what a lessor complaining of a failure to explore must prove. As has been discussed, to prove a breach of the covenant for reasonable development the lessor must show that the development well demanded has a probability of profit. If the same degree of probability is required to establish that an exploratory well should be drilled, then the covenant for further exploration is no different from the covenant for reasonable development. But requirement of proof of probable profit will impose a burden on lessors that can be met only in rare circumstances where the lessor's complaint is failure to explore. Historically, the oil and gas industry in the United States has found hydrocarbons in paying quantities with only about 1 in 10 exploratory wells, and only 1 of every 50 exploratory wells has discovered significant reserves. Therefore, it is unlikely that a lessor can prove a probability of profitability for an exploratory well.

Those commentators and cases that have rejected the covenant for further exploration as a separate covenant have done so on the premise that showing a probability of profit is the essence of the reasonable prudent operator standard. The reasoning of the Oklahoma Supreme Court in *Mitchell v. Amerada Hess Corp.*, 638 P.2d 441 (Okl.1981), is an example:

"Failure to recognize the profit motive as an instrumental force in oil and gas leases on behalf of both lessee and lessor is to ignore the very essence of the contract. . . . Meyers' formulation of the proposed implied covenant ignores the potential for profit. . . . Can the duties of the lessee be judged apart from the spectre of profit where the activity is judged exploration rather than development? To do so is unwise and unnecessary. . . . It is simply not realistic to ignore profit as a consideration of the standard of a prudent operator simply because the lessor demands a wildcat be drilled on a productive lease rather than an additional well to a productive formation." Id. at 447.

This reasoning may overstate the argument for the covenant for further exploration. Professor Meyers did not ignore profit potential. He identified a number of factors that bear on whether the operator acquitted himself properly in exploration. Included is the feasibility of further exploratory drilling, which takes into account economic factors like the presence of geologic formations likely to contain oil or gas, the costs of drilling, the market for the product, and the size of the block necessary to drill a test well.

Because of the reasonable prudent operator standard, the law should recognize an implied promise by the lessee to explore further as well as to develop reasonably. What will the reasonable prudent operator acting to maximize profit do with untested formations and tracts? If he drills only those wells that are more likely than not to be profitable, he will be out of business when known deposits are fully developed. What the reasonable prudent operator will do—and does do in the "real" world—in the interest of long run profitability, is to devote portions of his drilling budget to exploration as well as to development.

Of course, the potential for profitability is relevant. Exploratory wells are not drilled without a close evaluation of the likelihood that operations will be successful. Recognizing that the odds are against the success of any particular exploratory well, the reasonable prudent operator will take great care in choosing which exploratory prospects to drill. But drilling exploratory wells is a matter of survival in the long run. Therefore, an acceptable probability of profit is necessarily lower for exploratory wells than for development wells.

b. Elements of Proof

Where the implied covenant for further exploration is recognized as separate from the implied covenant for reasonable development, the lessor has the burden of proof of showing that (1) there is a reasonable expectation that additional exploration will be successful, and (2) that the lessor's operator is behaving imprudently by failing or refusing to explore further.

As has been discussed, proof of reasonable expectation of profitability from further exploration requires a lessor to prove a lesser probability of profit than that required for breach of the covenant for reasonable development. However, the burden of showing that the lessee is behaving imprudently in failing to explore is usually much heavier than the parallel proof of breach of the covenant for reasonable development. As the risk of unprofitability increases, so does the apparent reasonableness of a lessee who declines to act.

Furthermore, in the short to medium run, the covenant to further explore does not require actual drilling. Based as it is on the reasonable prudent operator standard, the implied covenant does not require that drilling be conducted where preliminary measures will decrease the risks of actual drilling. Geologic exploration or geophysical testing will satisfy the obligation. In the long run, however, actual drilling of every part of the leased premises, both vertically and horizontally is required. In the long run, a failure to actually drill gives rise to an inference that the lessee has no intention to actually drill on undeveloped portions of the lease. That, in turn, suggests a speculative motive in seeking to continue to hold unexplored portions.

c. Stumbling Blocks to Enforcement

(1) Notice to the Lessee

Because the implied covenant for further exploration and the implied covenant for reasonable development are closely related, the stumbling blocks to enforcement are much the same. The lessor who demands cancellation of his lease because of the lessee's failure to explore further must generally show that he has given the lessee notice of his demand and a reasonable time to comply. Notice is crucial because cancellation is the usual remedy sought.

(2) Disclaimer or Limitation in the Lease

Likewise, a lessor's right to demand further exploration may be limited either by specific provisions in the oil and gas lease or the implication from leased provisions. Indeed, because the specific actions required by the implied covenant for further exploration are so imprecise, any definition of the duty, however limited, is likely to be accepted as reasonable.

Yet, such limitations are rarely encountered. That may be because lessees who take leases intend to explore them and so are not concerned about disclaiming or limiting their obligations. It may also be because limitations are unacceptable in the market place.

d. Remedies for Breach

Where breach of the implied covenant for further exploration is established, the usual remedy awarded is partial cancellation or conditional cancellation of the lease as to the unexplored area. One reason that cancellation is the favored remedy is the uncertainty that there has been any actual damage to the lessor in terms of lost royalties as a result of failure to explore. Another is that the burden of proving that the operator has been imprudent in failing to explore is so heavy that where it is met the lessee's incompetence or speculative intent is usually so apparent that equity demands termination of the lease.

4.   THE IMPLIED COVENANT TO PROTECT AGAINST DRAINAGE

The courts have been quick to recognize an implied promise by the lessee to protect the leased premises against drainage. This covenant may obligate the lessee to act even though there has been no development and the lease is held by payment of delay rentals. The lessee has a right under the lease to develop or not as he may choose during the primary term, but his discretion does not extend to permitting drainage.

Modern drilling and spacing rules lessen substantially the need for the implied covenant to protect against drainage. If the size of spacing units has been properly established, there should be no drainage between leases. However, spacing unit configurations are merely approximations. They are usually established on the assumption that drainage will occur in a circle around the well. In fact, the structure of the producing formation may cause drainage in a very different configuration.

The covenant to protect against drainage usually involves alleged drainage directly from one tract to a well on a contiguous tract. However, there may be an obligation to protect against indirect or fieldwide drainage. In *Amoco Production Co. v. Alexander,* 622 S.W.2d 563 (Tex.1981), the lessors complained that Amoco had permitted drainage from their leases in the lower area of a tilted reservoir, while actively producing from leases in the upper region of the same reservoir.

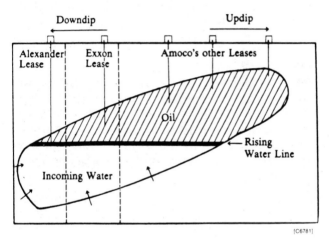

[C6781]

Alexander contended that Amoco had in fact accentuated the process by plugging wells on the Alexander leases and accelerating production from leases higher on the reservoir. The Texas Supreme Court held that Amoco had an obligation to protect the Alexanders' properties against both local and fieldwide drainage. The obligation might require the operator to drill additional or replacement wells, rework existing wells, or seek voluntary or compulsory unitization or other administrative relief to protect the lessors' interest.

a.   Elements of Proof: In General

Where the lessor shows (1) substantial drainage from the leased premises and (2) a probability that an offset well would be profitable, breach of the implied covenant to protect the premises against drainage will be found. As a general rule, the burden is upon the lessor to prove breach. By granting the lease, the lessor has granted to the lessee the right to make operating decisions. Unless the lessor proves that the lessee is acting imprudently, there is no reason to change the agreement of the parties. Furthermore, if drainage is actually taking place, the lessee has a bigger stake in protecting the property than the lessor since the lessor's share of production is generally larger than the lessor's royalty. The lessee's self-interest should protect the lessor.

(1) Substantial Drainage

It is unclear how much drainage is required to be "substantial." A few cases suggest that it must be "in paying quantities". It ought not be so large. The requirement of substantial drainage is to prevent a lessor from harassing his lessee by seeking development under the guise of complaining about drainage. Therefore, substantial drainage should be an amount large enough under the circumstances to be reasonably of concern to the lessor.

(2) Probability of Profit

Requirement that the lessor show a probability of profit from the demanded protective measures is the key to the implied covenant to protect against drainage because it embodies the reasonable prudent operator standard. The reasonable prudent operator will not act simply because there is drainage from the leased premises. A protection well will be drilled only where it appears that it will either prevent the drainage or compensate for it by counter-drainage and produce in quantities sufficient to repay the lessee's cost of drilling, completing and equipping the well, plus a reasonable profit. Without a probability of profit, the reasonable prudent operator will choose to minimize his losses by permitting the drainage.

b. Elements of Proof: Where There Is Drainage by the Lessee

A much discussed issue concerning the implied covenant to protect against drainage is whether a lessor's burden of proof should be modified where the drainage complained of is caused by wells on another property operated by his lessee. This is called the "common lessee" situation. For example, suppose that O leases his property to A Company, as does O's neighbor X. Subsequently,

A Company drills a well on X's property that O contends drains his property.

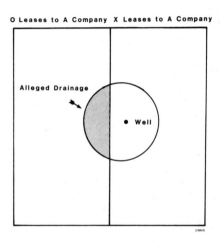

In such a situation, it is argued, O ought not have the burden of proving probability of profit because O is not protected by A Company's self interest. It will not matter to the common lessee if there is drainage from O's property. A Company will receive the working interest share of production whether O's property is drained from a well on O's property or from a well on X's property. Indeed, there may be an economic incentive for A Company to drain O's property from a well located on X's property, if O's lease provides for a higher royalty percentage than X's lease.

Many courts ignore the conceptual problem and impose the usual burden of proof upon the lessor. However, since breach is based upon a finding of

the facts of substantial drainage and probability
of profit, inferences of unfair dealing that may
arise from the operator's position as a common
lessee may affect the weight given the lessor's
evidence.

Some courts have gone further and suggested
that where there is a common lessee the lessor
should be relieved of the burden of proving
probability of profit. If the lessor can show that
there has been substantial drainage to a well on
premises operated by his lessee, then he is entitled to his remedy. The common lessee's inherent conflict of interest creates an inference of
bad faith where drainage occurs. Such cases
have sometimes seen the common lessee situation
as creating another implied covenant. In *Cook v.
El Paso Natural Gas Co.*, 560 F.2d 978 (10th Cir.
1977), the court recognized an implied covenant
of the common lessee to refrain from any action
that would injure his lessors' property.

Several commentators have urged that the requirement of showing a probability of profit
ought not be waived in the common lessee situation. The reasonable prudent operator standard
is meaningless without reference to probability of
profit. The lessor whose property is drained by a
well operated by his own lessee is in no worse position than he would be if his property were being
drained by some other operator. At most, the
procedural rules might be changed in the common
lease situation so that the burden on profitability
would be reversed; where the lessor showed that

operations of his lessee on another property were causing substantial drainage from his lease, the lessor would prevail unless the lessee was able to prove that protection probably would not be profitable.

c.   Stumbling Blocks to Enforcement

(1) Notice to the Lessee

As with other implied covenants, notice of the alleged breach and a reasonable period of time to correct the failure to act are usually prerequisites to the granting of the equitable remedy of cancellation.   However, an exception may be made where drainage occurs in a common lessee situation, on the theory that the lessee knew or should have known of the lessor's complaint.

(2) Disclaimer or Limitation in the Lease

The implied covenant to protect against drainage can also be disclaimed or limited by agreement of the parties, at least if it is implied in fact. Language is occasionally seen that if applied literally will obviate the implied covenant.   For example, consider a lease clause like the following:

"In the event a well or wells producing oil and gas in paying quantities should be brought in on adjacent land and within 150 feet of and draining the leased premises, lessee agrees to drill such offset wells as a reasonably prudent

[*304*]

operator would drill under the same or similar circumstances.''

Although such language may look at first glance like an express statement of the covenant to protect against drainage for reasonable development, the reference to distance may severely limit the covenant. Ordinarily, spacing rules will not permit a well to be drilled within 150 feet of a property line. Thus, the specific promise is in effect a disclaimer, for it obligates the lessee to protect the premises only in a situation that cannot legally arise.

The courts may refuse to enforce such a provision on a variety of grounds. It may be held overreaching or unconscionable. It may be held to be predicated upon a mutual mistake of fact concerning the spacing rules. It may be enforced against the lessor where the drainage results from the actions of some other lessee but not where the drainage results from the action of a common lessee. Or, it may be construed as a narrow affirmative statement that does not limit the implied covenant. However, the possibility that such provisions may be enforced leads many lessors to strike from leases any reference to implied covenants and to affirm specifically in the lease the lessee's obligation to act as a reasonable prudent operator. This approach is based on the premise (that is likely to be correct) that any reference to implied covenants in a lease is likely to be for the purpose of limitation rather than restatement or expansion.

(3) Waiver or Estoppel

A final stumbling block to the lessor in enforcing the implied covenant to protect against drainage is the possibility of waiver or estoppel by accepting delay rentals. Several courts have held that a lessor cannot complain of failure to protect the premises for any period of time for which he has accepted delay rental payments. The theory is that delay rental payments are payments for the privilege of maintaining the lease without drilling a well. As a result of such decisions, lessors who believe their property is being drained should reject delay rentals tendered.

Acceptance of delay rentals ought not bar a lessor from asserting breach of the implied covenant. As has been discussed, provision for delay rentals is included in leases by lessees to make clear the lessee's right to hold the lease during the primary term without obligation to test the premises. Lessors are willing to waive the covenant to test because they are compensated by bonus and rentals and their oil and gas remains in place. They do not intend that payment of delay rentals will give the lessee the right to let the property be drained.

d.   Remedies for Breach

(1) Damages

Damages are normally an adequate remedy for breach of the implied covenant to protect the premises against drainage. The lessor who is compensated for lost hydrocarbons is made whole. Some courts have measured damages by applying the lease royalty to the amount of drainage. Others have calculated royalty on production that could have been obtained from an offset well. Professors Williams and Meyers have suggested that measure of damages ought to be the amount of the drainage that could have been prevented by drilling an offset well. However, that logical limitation has generally been ignored.

(2) Cancellation or Conditional Cancellation

Lease cancellation or conditional cancellation is also available as a remedy for breach of the covenant to protect against drainage. Except in a few states, it is unlikely to be ordered unless the lessee has failed in his obligation of good faith and fair dealing toward the lessor or unless damages are impossible to ascertain.

5. THE IMPLIED COVENANT TO MARKET

The implied covenant to market imposes upon the lessee the duty to use due diligence to market oil and gas produced within a reasonable time and at a reasonable price. It follows from the fundamental premise that the reasonable prudent operator seeks to maximize profit. The implied covenant requires the lessee to use the diligence of a prudent business person in finding a market and negotiating a sale.

Problems with the implied covenant to market usually involve the sale of natural gas rather than oil. One reason is the difference in the physical characteristics of oil and gas. Oil is easily stored, and is sold on a "spot" basis from the storage tank in which it is placed after production. Because gas must be sold into pipelines, it usually is sold under complicated long term contracts. Delays of several years between completion of a gas well, negotiation of a gas contract, and extension of a pipeline to take the production are not unknown. Another reason is that most lease royalty clauses provide for payment of oil royalty in kind, so that the lessor can make his own arrangements for sale. In contrast, gas royalty is usually a percentage of the sale price received by the lessee.

a.   Within a Reasonable Time

How much time is reasonable will depend upon the facts and circumstances. Even if there is a ready purchaser for the production, it may be prudent to wait to get better terms or to deal with another pipeline; decline of demand for natural gas in times of economic recession does not fall equally upon all pipelines. Contract negotiation may be time consuming because of the usual long duration of agreements and regulatory requirements. Delays of several months are usual.

b.   At a Reasonable Price

The lessee's duty to market oil and gas at a reasonable price really means at the best possible price. Of course, price is merely the lessee's first priority. Other terms of the sale may have importance as well. But ordinarily, the reasonable prudent operator will seek to maximize profit by selling his product at the highest available price.

c.   Proof of Imprudence

Where the lessee is held to have breached his obligations, it is often because of failure to deal fairly with the lessor. *Amoco Production Co. v. First Baptist Church of Pyote*, 579 S.W.2d 280 (Tex.Civ.App.1979), writ ref. n.r.e. 611 S.W.2d 615 (1980), is a good example. There, Amoco entered into a long term gas sales contract in 1969. Market conditions at the time were poor, and in order to sell its production, Amoco had to commit to the contract future production from undeveloped leases. In 1973, a well was drilled on a 640 acre unit that included several of the previously undeveloped leases dedicated to the 1969 contract. By that time, gas prices had climbed substantially. Amoco negotiated an amendment to its 1969 contract by which additional leases in the 1973 unit were committed to the contract and, in return, the contract price for all gas sold by Amoco was increased. Though the renegotiated contract was obviously good business for Amoco, several lessors whose leases had been committed to the

[*309*]

1969 contract by the amendment sued alleging breach of the implied covenant to market. They presented facts showing that other owners had contracted to sell their gas in 1974 and 1975 on substantially better terms than those obtained by Amoco. The court found breach of the implied covenant, because Amoco had obtained a substantial improvement in the price that it received under the 1969 contract by trading off the interests of the lessors.

d. Remedies for Breach

(1) Failure to Market Within a Reasonable Time

Where the breach complained of is a failure to market production within a reasonable period of time, the lessor will ordinarily claim that the lease has ended. If the claim is brought during the primary term, the lessor will demand that the court exercise equity powers to cancel the lease. If the claim is made after the primary term, the lessor's theory will be that the lease has terminated by its own terms because of the lessee's failure to secure production in paying quantities.

Demand for cancellation or termination is not likely to fall on receptive ears. Once a lessee has gone to the risk and expenditure of developing a well on leased premises, the courts are reticent to find that his rights have terminated. As a general rule, damages will be the remedy granted to the lessor.

(2) Failure to Market at a Reasonable Price

It is particularly likely that damages will be the remedy assessed if the breach complained of is a sale by the lessee at less than a fair price. Where the lessor complains that his royalty is not calculated on an adequate price, he can be made whole by damages.

However, interesting problems may be encountered in calculating damages due. *Amoco Production Co. v. First Baptist Church of Pyote*, discussed above, is an example. There, the gas royalty clause provided that the lessor was to be paid royalty calculated on the basis of the "amount realized" from the sale of the natural gas. Having found that Amoco had marketed the production at less than a fair price, the court awarded the lessors damages based upon the fair market value of the gas when delivered. The court reasoned that since Amoco failed to obtain a "fair" price for the gas it marketed, the amount realized from the sale was inappropriate as a basis for royalty. The result was that Amoco's lessors were able to collect royalty for their share of the natural gas produced on a price higher than that received by anyone selling gas from the unit. The remedy fashioned reflects an attempt by the court to punish Amoco for its abuse. As such, it demonstrates the scope of courts' powers.

e. Stumbling Blocks to Enforcement

(1) Notice to the Lessee

Notice to the lessee of the claimed breach of the implied covenant to market is usually held essential where the lessor asks the court to exercise equitable powers to cancel the lease. Notice may also be required by the contract principle of cooperation wherever the basis for the claim of breach of the implied covenant is the lessee's failure to market the product within a reasonable period of time; presumably, the lessee can mitigate the lessor's damages by acting promptly after such a notice.

However, where the asserted breach is the lessee's failure to sell the production at a fair price, notice should not be a prerequisite. Most lease royalty clauses are structured so that the lessee owns all of the natural gas produced, with the lessor's royalty being based upon the price for which it is sold. By the time the lessor becomes aware of the sale, the production will have been committed to contract and there will be no purpose for the notice.

(2) Waiver or Estoppel

The lessor should not be held to have waived or be estopped from asserting the implied covenant to market unless he has knowingly ratified the terms of a gas contract. Merely accepting payments of royalties tendered should not establish grounds for waiver or estoppel because there is no question but that the amount of royalties tendered is due; the lessor's contention is that a larger amount is due. However, the indicated result may not hold where the lessor has executed a division order agreeing to accept the amount of royalty tendered as full payment, as has been discussed at pages 272–274.

(3) Disclaimer or Limitation in the Lease

Where the lessor's claim is that the lessee has not acted within a reasonable time to market production, it has been asserted that the presence of a shut-in royalty clause in the lease should bar suit. The lessee has the right under a shut-in royalty clause to maintain the lease without actual production and marketing by paying a shut-in royalty, as is discussed in Chapter 9. The argument is that the shut-in royalty clause gives the lessee the option of maintaining the lease either by production or by payment of shut-in royalties.

The argument has been rejected by the courts. The purpose of the shut-in royalty clause is to protect the lessee against loss of the lease for failure of "production" where marketing is not possible or advisable, not to relieve him of the duty to market. However, payment of shut-in royalty may bear on the reasonableness of marketing delays. Where a lessor has received shut-in royalty payments, his complaint that marketing has not occurred quickly enough is weakened.

[*313*]

6. THE IMPLIED COVENANT TO OPERATE WITH REASONABLE CARE AND DUE DILIGENCE

The implied covenant that the lessee will operate with reasonable care and due diligence overlaps several of the other implied covenants. For example, the decision in *Waseco Production v. Bayou States Oil Corp.*, discussed in conjunction with the implied covenant to reasonably develop at pages 285–286, could have been based on the implied covenant to operate with reasonable care and due diligence. The reasonable prudent operator is competent, and will use enhanced recovery techniques where they will be profitable. Likewise, the result reached by the court in *Amoco Production Co. v. Alexander*, discussed in conjunction with the implied covenant to protect against drainage at pages 299–300, may also be seen as an application of the implied covenant to operate with reasonable care and due diligence. The reasonable prudent operator will seek to protect both his interests and his lessor's interests by seeking voluntary or compulsory pooling or unitization or appropriate administrative action.

As Professors Williams and Meyers note in their treatise, complaints that the implied covenant to operate with reasonable care and due diligence has been breached usually are based upon one of four types of objections. They are (1) that the lessee has damaged the property by negligence or incompetence, (2) that the lessee has damaged the lessor by premature abandonment of a well capable of producing in paying quantities, (3) that the lessee has failed to use advanced production techniques, or (4) that the lessee has failed to protect the lessor by failing to seek favorable regulatory action.

It is important to note that this categorization does not limit the scope of the implied covenant. As has been discussed, all of the implied covenants may be seen as applications of the reasonable prudent operator standard. The courts have been quick and creative in extending implied covenants to protect lessors' interests. Because the implied covenant to operate with reasonable care and due diligence is the broadest of the six commonly encountered implied covenants, it is likely to be used by the courts to remedy future problems even though they do not fall clearly within the categories noted.

D. THE FUTURE OF IMPLIED COVENANTS

Some predict that implied covenants will become less important as lessors become more sophisticated and demand express covenants. Rapidly increasing oil and gas prices have caused steep increases in bonuses paid for leases and led lessors to pay much closer attention to lease terms than in the past. To the extent that oil and gas leases are not executed on preprinted forms but are specially negotiated by the parties to fit their particular circumstances, it may be expected that they will leave less room for application of implied covenants.

Professor Patrick H. Martin has made a forceful argument that the concept of the reasonable prudent operator should be expanded to take economic and social factors into consideration in addition to profitability. Professor Martin argues that the law of implied covenants has ill-served the nation because it has resulted in more development than is economically necessary to maximize production.

One should not expect to see either lesser emphasis on implied covenants or a shift in the focus of implied covenants. The importance of implied covenants and their focus upon the narrow relationship between lessor and lessee is due primarily to the case by case method by which problems come to the courts. When judges and juries are presented with facts demonstrating that lessees have acted incompetently or with speculative motives toward lessors, the existence and breach of implied covenants is likely to be found, usually with little concern about whether they are implied in fact or at law or whether they are consistent with broader public policy. Implied covenants are an available and effective way of redressing wrongs.

CHAPTER 12

LEASE TRANSFERS

Lease interests are frequently transferred. Leases are viewed by oil companies as inventory, and are frequently traded to permit development of lease "blocks." Fractional working interests are often assigned to investors who put up the money for drilling. Overriding royalty or other nonoperating interests may be transferred to the landman who acquired the lease, the geologist who developed the prospect, or to others who are instrumental in structuring the venture. In this chapter, we will look at common problems presented by transfers of lease interests.

A. LESSEE'S RIGHT TO TRANSFER

Interests under oil and gas leases generally are treated by the courts like interests in real property. They are freely assignable unless reasonable limitations are imposed by the terms of their creation. Therefore, even without provisions in the lease permitting assignment, a lessee may transfer all or any part of his interest.

Most lease forms contain language that specifically permits assignment. An example is the first sentence of paragraph 8 of the two leases in the Appendix: "The rights of Lessor and Lessee may be assigned in whole or in part . . . ." A specific provision permitting assignment is included because the nature of the lessee's interest under an oil and gas lease has been described in a variety of ways ranging from a mere license to a fee simple determinable interest in minerals. If common law doctrines were applied strictly, there would be no implied right to divide the lease in states where the leasehold interest is classified as a profit in gross; though profits in gross were assignable at common law, they were not divisible.

B. EFFECT OF TRANSFER ON THE LESSOR

1. FURTHER RIGHTS AGAINST THE LESSEE

One may assign his rights under a contract but he remains bound by his obligations because of privity of contract. Therefore, unless the lease provides otherwise, a lessee who assigns his operating rights under a lease remains liable for future breaches. The lessee can protect himself by imposing a specific obligation upon the assignee to indemnify him against such claims. More generally, lessees seek to change the general rule by provisions in the lease. Language from a lease used in North Dakota is typical:

> "In case Lessee assigns this lease, in whole or in part, Lessee shall be relieved of all obligations with respect to the assigned portion or portions upon furnishing the Lessor with a written transfer or assignment or a true copy thereof."

Similar language is found in paragraph 8 of the Texas and Colorado leases included in the Appendix. Though the exculpatory terms are not restricted to breaches occurring after the transfer, they are so limited by the courts.

2. FUTURE RIGHTS AGAINST THE TRANSFEREE

a. Lease Covenants Run With the Land

The transferee of a lessee's interest under an ordinary real property lease is bound by provisions that "run with the land." Covenants are held to run with the land where (1) they were intended to do so by the original parties, (2) they pertain to matters that "touch and concern" the land, and (3) there is privity of estate between the original lessor and the transferee. Where these elements are present, transferees of leasehold interests are held to be bound by both express and implied provisions of the lease.

Almost without exception, the provisions of oil and gas leases have been held to be covenants that run with the land. Therefore, the transferee of a leasehold interest assumes all of the obligations imposed by the original lease. Of course, the assignment may impose additional obligations upon the transferee. But nothing in the assignment can limit the rights of the original lessor against the transferee under the terms of the lease.

b. The Assignment/Sublease Distinction

The obligations of the transferee to the lessor may be affected by whether the transfer is an assignment or a sublease. An assignment is a transfer of the entire leasehold interest in all or any portion of the property. A sublease is a transfer of less than all of the interest. Where the transaction is classified as an assignment, the lessor can enforce covenants of the lease that run with the land against the transferee as well as his lessee; the lessor will be in privity of estate with the transferee and in privity of contract with the lessee. If the transaction is classified as a sublease, the lessor's rights are against the lessee/sublessor and the transferee's obligations are owed to the lessee/sublessor, because there is no privity of estate between the lessor and the sublessee.

Some jurisdictions have applied the assignment/sublease distinction to transfers of oil and gas leases. Where the lessee has assigned the lease for less than the entire remaining term, or reserving a right to reenter or an overriding royalty interest, it has been held that a sublease is created rather than an assignment. Louisiana recognizes the distinction routinely. However, the distinction has been largely ignored because it is not recognized by the customs and usages of the industry.

C. RIGHTS AND DUTIES OF THE LESSEE AND HIS TRANSFEREE

Where the lessee retains rights after the transfer of the lease, whether to a non-operating interest or an operating interest in a portion of the leased premises, disputes frequently arise over the relationship of the transferor and transferee.

1. PROTECTION OF NON–OPERATING INTERESTS BY IMPLIED COVENANTS

Authority is divided whether a lessee who assigns his interest retaining an overriding royalty or other non-operating interest is entitled to the protection of implied covenants. Most of the older cases suggest that he is not. More recent decisions often assume that he is.

Non-operating interests should be protected by implied covenants just as are lessors. With assignments as with leases, it is likely that a substantial part of the consideration for the transfer is the expectation of profits from production. In both situations, the transferor will often not be sophisticated enough to draft specific contractual protection for himself. The protection of the usual lease implied covenants can be implied either from the facts of the transaction or in law in the interest of fairness.

Extension of the protection of implied covenants to the lessee who transfers his operating rights in exchange for a non-operating interest will not materially affect the obligations of the holder of the lease. The lessee's transferee will have no obligation to test the premises, unless there is a specific drilling obligation imposed by the assignment. The transferee will ordinarily have no liability to the original lessee for letting the lease terminate by failing to pay delay rentals. The transferee will merely have an obligation to the lessee as well as to the lessor to protect the premises against drainage, to reasonably develop and explore once production is obtained, and to do all the other things that a reasonable prudent operator would do under the circumstances.

2.   PROTECTION OF NON–OPERATING INTERESTS AGAINST "WASH OUT"

A frequent problem after a lessee has transferred his operating rights in a lease and retained a non-operating interest is the "wash out." If the transferee permits the lease to terminate and then subsequently re-leases the property, should the original lessee's non-operating interests be recognized under the new lease? Not affording him such protection will tempt transferees to unfairly maximize profits by washing out non-operating interests. On the other hand, requiring recognition of the right in lease extensions or replacement leases will hamstring transferees; there may be sound business reasons to permit a lease to terminate and then re-lease the property. The transferee should not be a fiduciary for the transferor.

Most cases have held that one who transfers operating rights retaining a non-operating interest is not protected by implied covenants against wash out. However, a few cases have extended the transferor protection on the grounds that either (a) a constructive trust is created by the close relationship between the parties as shown by the particular facts, or (b) the facts give rise to an inference of bad faith by the transferee.

As a result of the uncertainty whether a lessee who transfers his operating rights in a lease will be protected against a washout and when such protection will be extended, lease assignments reserving non-operating interests frequently contain specific provisions guaranteeing recognition of the transferor's non-operating interest in lease extensions and renewals. An alternative is to obligate the transferee to offer to reassign the lease before permitting it to terminate. Either type of provision should protect the interests of the transferor and obviate any implied obligation.

3. IMPLIED COVENANTS OF TITLE FROM THE LESSEE

Is the transferee of a lease entitled to protection from covenants of warranty implied in the assignment from the lessee? The transferee is protected by the covenant of warranty of the lessor found in most oil and gas lease forms. But implied covenants of title from the lessee are generally held not to be created by the language of an assignment. However, the language of assignments may differ substantially from case to case, and in some states the use of particular words of grant may imply certain covenants. Therefore, many assignments of leases specify whether covenants of title are intended.

4. PERFORMANCE OR FAILURE OF PERFORMANCE BY A PARTIAL ASSIGNEE

Frequently disputes arise as to the effect of the transferee's actions or failure to act upon the rights of a lessee who partially assigns operating rights. Suppose, for example, that A Company takes a lease on a 640 acre section of land and subsequently assigns the east $\frac{1}{2}$ of the section to B Company:

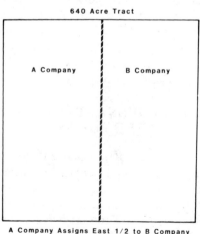

640 Acre Tract

A Company   B Company

A Company Assigns East 1/2 to B Company

What is the effect of production by B Company? What is the effect of B Company's failure to pay delay rentals?

Both situations are governed by the general principle that the rights and duties of the lessor and the lessee under the lease are set when the lease is originally granted. The term clause of the lease provides that the lease will extend so long as there is production "from the leased premises." If B Company obtains production on the property, the lease is extended for both the west and the east portions. Even where the term clause of the lease requires "production by the lessee," the courts have generally held that the language is satisfied if production is obtained by the lessee's assignee or one to whom the lessee has agreed to assign. On occasion, there has not even been a written agreement to assign.

Likewise, unless the lease provides otherwise, a failure either by A Company or B Company to make proper payment of delay rentals will cause the lease to terminate upon both the east and west portions of the tract. The lease calls for delay rental payments to be made covering the whole premises described. Most modern oil and gas lease forms contain specific language to change this result. A typical example is the following from a New Mexico lease form:

"In the event of an assignment of this Lease as to a segregated portion of said land, the rentals payable shall be apportioned as between the several leasehold owners ratably according to

[*325*]

the surface area of each and default in rental payment by one shall not affect the rights of other Leasehold owners hereunder."

The effect of the language is to make delay rentals divisible where the lease is subdivided by the lessee.

In Louisiana, where the lease contains a clause like that quoted, A Company would be required to establish production on the west half in order to satisfy the term clause as to that portion, even if B Company had obtained production on the east half. The Louisiana courts have held that the clause makes the lease divisible, so that each portion must be treated as a separate lease.

5. DIVISIBILITY OF IMPLIED COVENANTS AFTER A PARTIAL ASSIGNMENT

An exception to the general rule that lease obligations are not divisible is recognized in some states after a partial assignment with respect to the implied covenants to reasonably develop and to explore further. At issue is whether the obligation of the lessee and his assignee is to be judged by reference to the lease as a whole or whether each must stand on its own. As has been discussed in the example above, production from either portion of the subdivided lease satisfies the term clause of the lease for the whole property. Since the lease is not divisible, both A Company and B Company have an obligation to continue to develop reasonably. But should the diligence of the lessee and his assignee be judged on the basis of the lease as a whole or separately for the two portions? Logic and adherence to the principle that rights and obligations under the lease are set when it is formed indicate that the lease should be considered to be indivisible for the purposes of implied covenants just as it is for express covenants. However, several cases have specifically rejected that conclusion, and others are unclear.

PART IV

TAX AND BUSINESS MATTERS

CHAPTER 13

OIL AND GAS TAXATION

Oil and gas taxation is usually covered in a course separate from the basic course in oil and gas law. However, an understanding of the fundamental rules of oil and gas taxation is essential for any lawyer practicing oil and gas law and for any landman or manager in the industry; routine transactions may have important tax implications. This chapter will consider taxation of oil and gas operations on a transactional basis.

A. BASIC PRINCIPLES

In most respects, taxation of oil and gas transactions is consistent with general tax principles, which are beyond the scope of this book. However, there are at least three concepts that are peculiar to oil and gas taxation and of fundamental importance to oil and gas operations. They are (1) the property concept, (2) the intangible drilling cost deduction, and (3) depletion.

1. THE PROPERTY CONCEPT

For oil and gas taxation purposes, the term "property" refers to each separate legal interest owned by a taxpayer in each geological deposit, in each separate tract or parcel of land. Treas. Reg. § 1.614–1(a). Thus, the property concept has legal, geological, and geographical aspects.

The property concept affects oil and gas transactions in several important ways:

a. geological and geophysical costs incurred in the search for oil and gas properties must be allocated to a property unit and included as a part of the depletable basis of that property unit;

b. intangible drilling costs are deducted on a property basis and may have to be recaptured when that property is sold or disposed of;

c. depletion, whether percentage depletion or cost depletion, must be calculated separately for each property;

d. gain or loss from sales must be computed separately for each property.

2. INTANGIBLE DRILLING COSTS DEDUCTION

The deduction for intangible drilling costs is an important tax benefit under the Internal Revenue Code. Intangible drilling costs are those costs that have no salvage value and that are incurred incident to and necessary for the drilling of wells and their preparation for production of oil and gas.   Treas.Reg. § 1.612–4.

Were it not for a special provision of the Internal Revenue Code, intangible drilling costs would be capitalized with costs of equipment and depreciated over the lifetime of the well.   However, subject to strict limitation, § 263(c) of the Internal Revenue Code permits acceleration of the process and deduction of intangible drilling costs in the year when they are incurred.   The owner of a working interest in an oil and gas property has the option either to deduct intangible drilling costs as expenses in the year incurred or to capitalize them and recover them through cost depletion or depreciation.

Most taxpayers elect to deduct intangible drilling costs currently, on the theory that it is better to take tax write offs sooner rather than later. However, the benefit is not without a price.   Intangible drilling costs deducted in excess of cost depletion are an item of tax preference under § 57(a)(11) of the Code and may trigger minimum tax consequences.   Also, they may be subject to recapture under § 1254 when the property is sold or transferred.

An important limitation on the option to "expense" intangible drilling costs is the requirement that a taxpayer must both (a) incur the costs, and (b) own the full share of the working interest for which intangible drilling costs are deducted for the entire payout of the cost of drilling, equipping, and operating the well. This limitation is often referred to as the "complete payout concept." Oil and gas transactions are carefully structured to comply with or avoid the complete payout concept because of the significance of intangible drilling cost deductions. A common avoidance technique for a partnership is a special allocation of intangible drilling costs to the partners who contributed drilling funds. A special allocation will be upheld if it has "substantial economic effect", which requires that there be business "logic" to the allocation.

3.   DEPLETION

As a mineral deposit is produced, it is depleted. In a very real sense, production consumes the taxpayer's capital. Some part of every dollar received is in reality a return of a portion of the costs expended in developing the property rather than a true profit. Therefore, a deduction from current income for mineral depletion is appropriate.

The Internal Revenue Code recognizes two methods for taxpayers to compute depletion: cost depletion and percentage depletion. The taxpayer is required to make both calculations and to deduct the larger of the two amounts.

a.   Cost Depletion

Under cost depletion, the taxpayer in an oil and gas property deducts his basis in the property from his income as oil and gas are produced and sold. Cost depletion is calculated by a formula set forth in Treas.Reg. § 1.611–2 that can be expressed as follows:

$$B \times \frac{S}{(U+S)}$$

B   equals the adjusted basis of the property at the end of the period.

U   equals units remaining at the end of the period.

S   equals units sold during the period.

This formula relates the recovery of the taxpayer's investment to the proportion that the current unit sales of oil and gas bear to the total anticipated sales of oil and gas from the property. His investment is recovered ratably over the life of the reserves.

b.   Percentage Depletion

Percentage depletion is a statutory provision that permits deduction of specified percentages of the gross oil and gas income from a property in lieu of depleting the actual basis. A fixed percentage of each dollar received from production is excluded from taxation. Percentage depletion is not limited to the actual capital costs incurred by the taxpayer. Deductions for percentage depletion from a prolifically producing property may greatly exceed the actual capital costs incurred by the taxpayer in developing the property.

(1) Economic Interest Concept

An important limitation on the availability of percentage depletion is the requirement that the taxpayer own an economic interest in the property in order to claim it. The rationale of the limitation is that percentage depletion is intended as an incentive for those who own mineral deposits. However, as the term has been defined, ownership of the mineral interest itself is not required. Treas. Reg. § 1.611–1(b) provides that an economic interest is held where a taxpayer acquires any interest in minerals in place and secures, by any form of legal relationship, income from extraction to which he must look for return of his capital. Royalty interests, net profits interests, and carried interests are all generally recognized as economic interests.

(2) Independent Producers' and Royalty Owners' Limitation

Percentage depletion was originally enacted as a special incentive to promote oil and gas production. For approximately forty years it was broadly available and usually more advantageous for taxpayers than cost depletion. Ironically, in 1975 at the height of an oil shortage, the Tax Reduction Act of 1975 repealed percentage depletion for oil and gas produced after December 31, 1974. However, exceptions to repeal were provided in § 613A, one of which is for independent producers and royalty owners. For example, in 1983, taxpayers qualifying as independent producers or royalty owners are permitted percentage depletion at a 16% rate for a total of 1,000 barrels of oil or 6 million cubic feet of natural gas *per day* to the extent of 50% of the taxable income from the property and, with certain adjustments, 65% of the taxpayer's taxable income for the year. After 1983, the rate will be 15%. Thus, for individual investors and small oil companies, percentage depletion is still an important incentive.

B.   TAXATION OF TRANSFERS OF MINERAL RIGHTS

1.   LEASE PAYMENTS

a.   Lease Bonus

An oil and gas lease is usually granted for a lease bonus, a payment from the lessee to the lessor to induce the grant. Although a typical oil and gas lease conveys the lessor's mineral rights to the lessee, the bonus payment is taxable as ordinary income to the lessor. Prior to the Tax Reduction Act of 1975, the lessor could claim percentage depletion on bonus payments. Since then, the Internal Revenue Service has taken the position that bonus payments, though in the nature of advance royalties, are subject only to cost depletion unless there is actual production in the year of payment. As this is written, the courts are split.

The lessee must capitalize bonus payments as a part of geological and geophysical costs, which are discussed at page 342.

b.   Delay Rentals

Delay rental payments made by the lessee to the lessor during the primary term of a lease, are treated as rents from the land. As such, they are taxed to the lessor as ordinary income not subject to percentage depletion. A lessee who pays delay rentals may deduct them as an operating expense. He may also elect to capitalize them and recover them through cost depletion.

c.   Royalty Payments

Payments of royalties to the lessor will be taxed as ordinary income. Where they are based on actual production, they ordinarily qualify for the depletion allowance. The lessee excludes the value of production paid in kind as royalty income from his income. Where the royalty is paid as a percentage of the proceeds from the sale of production or where it is not based upon actual production, the lessee treats the royalty as an operating expense.

2.   LEASE TRANSFERS

a.   The Sale/Sublease Distinction

An important distinction for oil and gas taxation purposes is made between a sale and a sub-

lease.   Where the transaction is termed a sale,
the seller computes gain or loss on the difference
between the fair market value of the consideration received and his basis in the property transferred.   Since oil and gas properties are generally
treated as § 1231 assets, any gain on the property will be treated as a capital gain (long term, if it
has been held for the requisite period), while any
loss will be treated as an ordinary loss.   In contrast, if the transaction is termed a sublease, all
consideration received by the seller will be treated as ordinary income.   Therefore, classification
as a "sale" is more beneficial to the seller than
classification as a "sublease" because (1) the gain
may qualify for lower tax rates as long term capital gain and (2) the seller's basis may be deducted
in computing the gain.

What is the difference between a sale and a
sublease?   For tax purposes, a sale results when
(1) the seller transfers *all* of his interest or a
fractional interest that is identical in kind (i.e.,
working interest, royalty interest, etc.) to that he
retains, or (2) when the interest conveyed is of a
continuing non-operating nature (e.g., an overriding royalty interest conveyed from a working interest), or (3) when the seller conveys the working interest and retains only non-operating rights
of a non-continuing nature (such as a production
payment).   Any other type of transaction is
termed a sublease.

The sale/sublease distinction is particularly
troublesome because leases are often acquired

for speculative purposes by persons who then trade them to oil companies in return for cash consideration and a retained overriding royalty interest. Though the parties to the transactions are often not aware of it, the proper tax result is that all of the cash received by the transferor is taxable as ordinary income. The theory is that, because the transaction is a sublease, the basis of the speculator is accumulated in the overriding royalty interest retained. Some believe that the result may be avoided by retaining a fraction of the working interest in addition to an overriding royalty interest, by retaining a carried or working interest in place of an overriding royalty interest, or by transferring the lease to a partnership composed of the speculator and the assignee, and providing for benefits under the partnership agreement similar to those of an overriding royalty interest.

b.   Production Payments

Sometimes a lease will be transferred in exchange for a production payment. A production payment is a share of production from the leased premises that terminates when a specified amount has been recovered. An example would be "¹/₅ of all oil and gas produced and saved from said land, free of cost, until the market value at the well of such production shall aggregate One Million Dollars." For tax purposes, a production payment is treated as a mortgage. Thus, the transaction is treated as a sale. The transferor recognizes ordinary income or capital gain as the production payment is made. The transferee of the interest capitalizes the amount of the production payment. However, special rules may apply where the production payment is pledged for development.

c.   Sharing Arrangements

When one transfers an interest in a mineral property to another in exchange for a contribution from the transferee to the cost of acquisition, exploration, or development of the property, a sharing arrangement is generally created for tax purposes. The parties to the sharing arrangement are held to have "pooled" their capital and created an informal partnership for tax purposes. As a result, there are no tax consequences of the transfer until the property is abandoned or transferred, at which time the transaction will generally be taxed as a sale. See generally, G.C.M. 22730, 1941–1 C.B. 214.

A farmout, a transfer of an interest in acreage in return for drilling and testing operations on that acreage, is one of the most important sharing agreements. Rev.Rul. 77–176, 1977–1 C.B. 77 establishes an important limitation on such transactions. There, the Internal Revenue Service took the position that transfer of acreage in addition to that included in the drilling unit was a transfer of separate property not protected from taxation by the pool of capital doctrine. Worse, it said that the value of the property for tax purposes should be determined as of the date of the transfer, which usually is after the property has been proved by successful drilling operations. The revenue ruling limits the flexibility of sharing arrangements, and has spawned a variety of creative avoidance devices.

Bottom hole and dry hole contributions differ from usual sharing arrangements in that the party drilling does so on his own land and the party contributing support receives geological information to help him evaluate his own property, instead of an interest in the property being developed. Therefore, the Internal Revenue Service treats such transactions as sales rather than a sharing arrangement. The recipient of the funds recognizes income in the amount received. The contributing party must capitalize his expenditure as a geological and geophysical cost. See Rev. Rul. 80–153, 1980–1 C.B. 10.

C. TAXATION OF OIL AND GAS DEVELOPMENT

1. FORMS OF OWNERSHIP

Because of its high risk and large capital requirements, oil and gas exploration and development are often conducted by more than one person or entity. Problems then arise as to what form the joint ownership should take.

a.   Corporation

A corporate entity is attractive for oil and gas development because of the limited liability it provides. However, as a tax vehicle, a corporation is generally unattractive. Tax benefits in the form of accelerated deductions (such as intangible drilling cost deductions) or unrecognized income (such as that afforded by percentage depletion) are recognized at the corporate level, and do not pass through to the individuals or entities who own a corporation. Moreover, income earned by a corporation and distributed to its shareholders is subjected to double taxation, once at the corporate level and again as a dividend upon distribution.

Corporations that have elected to be taxed as partnerships under Subchapter S of the Internal Revenue Code have generally not been used as vehicles for oil and gas development because the technical rules applied have conflicted with important tax incentives. Percentage depletion in excess of cost depletion has created corporate earnings and profits. Distributions in excess of taxable income, computed with depletion, have been taxable as dividends, effectively generating a tax on the depletion deduction. However, the Subchapter S Revision Act of 1982 contained provisions designed to mitigate some of these problems. Subchapter S corporations may become a more popular form for ownership of oil and gas properties in the years ahead.

[*340*]

b. Partnership

A partnership is the ideal entity for oil and gas development from a tax viewpoint, because partnerships are not taxed under the Internal Revenue Code. Both profits and losses flow through the partnership to the individual partners. However, from a liability viewpoint, a partnership is a poor choice. Each partner is liable to the full extent of his assets for the torts and contracts of all of his partners. For this reason, partnerships are not usually used in oil and gas development.

Among small investors, limited partnerships are the preferred form of entity for oil and gas development. They offer limited partners the guarantee of liability limited to their actual investment. However, for tax purposes, they are treated like general partnerships. Limited partnerships are generally not favored by active participants in the oil and gas industry because limited partners are not permitted to participate in management.

c. Concurrent Ownership

A favored form of entity for oil and gas operations among those active in the oil and gas industry is concurrent ownership. Where parties jointly own and operate properties, each owns a separate property for tax purposes, and the tax results are approximately the same as if they were partners. On the other hand, liability of the non-operating party is effectively limited to the amount of his investment; the operator is treated as an independent contractor for his torts and contractual obligations. However, there are important and apparently growing exceptions to this rule that impose liability upon non-operators by strict liability theory or by statute.

A concurrent ownership arrangement may also be taxed like a corporation if it has more corporate than non-corporate characteristics. The rules are set out in Treas.Reg. § 301.7701–2(a)(1). Traditionally, concurrent ownership arrangements for production of oil and gas have maintained their non-corporate tax status by including in the operating agreements drafted to govern the relationship a specific recognition of the right of each owner to take his production in kind and limitations upon the right of the operator to commit unclaimed production. See I.T. 3930, 1948–2 C.B. 126; I.T. 3933, 1948–2 C.B. 130; and I.T. 3948, 1949–1 C.B. 161.

2. TAXATION OF SEARCH COSTS: GEO-LOGICAL AND GEOPHYSICAL COSTS

Expenditures made to obtain data to serve as a basis for the acquisition or retention of oil and gas properties are called geological and geophysical costs. Such costs include costs of core drilling and seismographic surveys. Geological and geophysical costs must be capitalized and recovered by depletion, like leasehold costs. Substantial accounting problems are presented in allocating geological and geophysical costs among properties.

3. TAXATION OF DEVELOPMENT COSTS

a. Intangible Drilling Costs

For tax purposes, costs incurred in preparing a drill site and actually drilling a well can be broken into two classes: intangible drilling costs and equipment costs. Intangible drilling costs are costs incurred that have no salvage value and are incurred incident to and necessary for the drilling and preparation of wells for the production of oil and gas. Typical intangible drilling costs expenditures include wages, fuel, services, and supplies used in preparing the drill site, drilling the well, and certain completion costs. However, intangible drilling costs do not include the costs of installing equipment for cleaning, processing or storage.

b. Equipment Costs

Oil and gas producing equipment is treated for tax purposes like the equipment of any other industry. The cost of such equipment must be capitalized and depreciated. Since 1981, accelerated depreciation under the Accelerated Cost Recovery System has been available, subject to recapture upon early disposition. Producing equipment will ordinarily qualify for investment tax credits, as well.

4. · TAXATION OF OIL AND GAS PRODUCTION

a.   Tax Treatment of the Owners of Production

The value of oil and gas production is taxed as ordinary income to its owners, subject to deductions for cost or percentage depletion.

b.   The Windfall Profit Tax

An excise tax is imposed upon "windfall profit" from crude oil removed from premises in the United States after February 29, 1980, by the Windfall Profit Tax Act. The term "windfall profit" is defined as the excess of the removal price of a barrel of crude oil over the sum of (1) the adjusted base price of a barrel and (2) the amount of the severance tax adjustment (if any) for a barrel.

The adjusted base price for a barrel of oil and the tax rate applicable is determined by a tier classification that may be summarized as follows:

Tier 1—oil from a property that produced before January 1, 1979.

Tier 2—oil from stripper wells (wells that produce ten barrels a day or less) or oil from an economic interest in a National Petroleum Reserve.

Tier 3—newly discovered oil (oil from an onshore property that did not produce in 1978 and oil from an offshore property leased after December 31, 1978 from which there was no production in 1978), heavy oil, and incremental tertiary production oil.

The tax rates on Tier 1 and Tier 2 oil are 70% and 60% of the windfall profit, respectively. The tax rate on Tier 3 oil was 30% through 1980, and is structured to "step down" from 27.5% in 1982 to 15% in 1986 and after as an incentive to new oil exploration. Independent producers (those who are not retailers or refiners) are given special treatment. Windfall profit on the first 1,000 barrels a day of Tier 1 and Tier 2 oil they produce is taxed at 50% and 30% respectively. Stripper oil is generally exempt after December 31, 1982. Royalty owners are entitled to a limited exemption for specified amounts of royalty production. For 1982 through 1984 the exemption is 2 barrels of oil a day. Starting in 1985, the exemption will be 3 barrels a day.

Because the Windfall Profit Tax is an excise tax, it is deductible in computing income taxes. For purposes of computing percentage depletion, gross income from a property is not reduced by the windfall profit. Therefore, the impact of the Windfall Profit Tax is ordinarily less than the high percentages suggest.

CHAPTER 14

OIL AND GAS CONTRACTS

Although oil and gas contracts are beyond the scope of most courses in oil and gas law, anyone who works in or with the oil and gas industry should be familiar with the types of contracts used in operations. In addition to the oil and gas lease, which has been considered in detail, these include:

Support Agreements

Farmout Agreements

Drilling Contracts

Operating Agreements

Gas Contracts

Gas Balancing Agreements

Division Orders

This chapter will provide a functional overview of these commonly encountered agreements.

All oil and gas agreements are subject to the general rules that govern contracts. A promise becomes a "contract," a legally binding agreement, only if there has been an offer and acceptance of clear and unambiguous terms pertaining to a subject matter that is not illegal or contrary to public policy, supported by consideration, between two persons or entities with the legal capacity to contract. Generally, oil and gas contracts must meet all the requirements for binding legal agreements.

Most oil and gas agreements are of sufficient duration or involve transfer of an interest in land, so that they are required by the Statute of Frauds to be in writing. Because of the large amounts of money at stake, almost all oil and gas contracts are in writing, regardless of the legal requirements.

A. SUPPORT AGREEMENTS

Support Agreements are contracts used to encourage and "support" drilling operations. Sometimes such agreements are called *contribution agreements*.

1. PURPOSE

In a support agreement, the contributing party agrees to contribute money or property in exchange for information. From the viewpoint of the contributing party, a support agreement is a purchase of geological or technological information. From the viewpoint of the party receiving the support, the support agreement lessens the cost or the risk of drilling operations.

2. KINDS OF SUPPORT AGREEMENTS

There are three commonly encountered kinds of support agreements. They are:

Dry Hole Agreement—the contributing party agrees to make a cash contribution if a dry hole is drilled. The party who may receive the money generally agrees to provide geological and drilling information whether or not the well is a dry hole.

Bottom Hole Contribution Agreement—the contributing party agrees to make a cash contribution in exchange for geological or drilling information if a well is drilled to an agreed depth.

Acreage Contribution Agreement—the contributing party agrees to contribute leases or interests in the area of the test well in exchange for information, if a well is drilled to an agreed depth.

With each of these agreements, it is essential that the parties clearly agree (1) what test information must be provided by the drilling party and (2) what conditions must be met in order for the contribution to be earned. Frequent disputes arise as to how dry a well must be in order to be a "dry hole" or what happens if the drilling party is unable to reach total depth agreed in a bottom hole agreement.

B. FARMOUT AGREEMENTS

A farmout agreement is an agreement to assign an interest in acreage in return for drilling or testing operations on that acreage. From the viewpoint of the *farmor,* the person or entity making the assignment, it may be used to maintain a lease by securing production close to the end of its primary term, to comply with an implied covenant to develop or offset, or to obtain an interest in production without additional cost. From the viewpoint of the *farmee*, the allure of a farmout agreement is that the farmee may earn acreage not otherwise available or at a lower cost than otherwise possible, or may be able to keep people and equipment gainfully employed.

A farmout agreement is closely related to support agreements. It supports oil and gas operations by spreading risks, costs and information. It differs from the support agreements discussed above in that the earning party conducts operations on the property of the contributing party rather than on his own property, and the parties end up sharing ownership of developed property.

The most important substantive issues of farmout agreements generally fall into two classes. They are:

(1) whether drilling the well is a covenant or a condition; and

(2) what must be done in order to earn the right to the assignment.

The distinction between a drilling *covenant* and a drilling *condition* relates to the liability of the farmee for failure to drill. If the farmee promises to drill a well on the premises, then the farmee may be held liable in damages for the reasonable costs of drilling if he fails to perform his promise. On the other hand, if drilling the test well is a condition of earning rights under the farmout

[*349*]

agreement, then there is no liability for failure to drill. The farmee simply earns no right under the terms of the farmout agreement. Often, farmout agreements make initial drilling a covenant of the farmee but permit the farmee to escape the obligation if unforseen conditions are encountered in drilling.

The terms of the farmout agreement relating to what the farmee must do to earn his rights determine the time and costs the farmee must incur. Farmout agreements usually require strict compliance by the farmee; that is, the well must be drilled to the full depth specified, all tests specified must be completed, and production sufficient to repay the costs of drilling and return a reasonable profit must be obtained. Terms are negotiable, however. Frequently, provisions for partial performance or substantial performance or earning without commercial production are encountered.

C. OPERATING AGREEMENTS

An operating agreement is a contract between cotenants or separate owners of oil and gas properties being jointly operated, that sets forth the parties' agreement with respect to initial drilling, further development, operations and accounting.

1. PURPOSE

An operating agreement defines the rights and duties of co-owners of oil and gas properties. It pools the jointly held fractional interests of the parties for operating purposes.

Most operating agreements are structured for use by cotenants. There are three "model" forms developed by the American Association of Petroleum Landmen in 1956, 1977, and 1982. Occasionally in the U.S. and often in Canada, one will encounter the model form developed by the Canadian Association of Petroleum Landmen. The provisions of the model forms are designed for use in transactions involving companies or persons active in the oil and gas industry. Frequently, the terms of the model agreements are modified substantially.

2. COMMON SUBSTANTIVE ISSUES

Although there are substantial differences in the provisions of operating agreements, all operating agreements must address common substantive issues in order to accomplish their basic purpose. There are at least three essential issues: (1) the scope of the operator's authority, (2) provision for initial drilling, and (3) additional development.

a. Scope of the Operator's Authority

Someone must be appointed as the operator and empowered by the co-owners to be responsible for operations on a day to day basis. The model form operating agreements provide for a narrow scope of liability for the operator ("gross negligence or willful misconduct") and limited basis for removal. The reason is that in dealings between companies and persons active in the industry, the operator usually acts as operator as an accommodation; perhaps because he has other operations in the same area. In contrast, many operating agreements cover the relationship of groups of investors with the operator who sold them their interests in the oil and gas property. The operator expects to make a profit from his operations and charges an appropriate fee. In such circumstances, the operator's investors may consider the limiting provisions of the model agreements overly restrictive.

Generally, operating agreements spell out in great detail both the obligations of the operator and limitations upon his discretion. Typically, duties are imposed upon the operator to carry certain amounts of insurance, maintain certain kinds of accounts, and hire legal and accounting assistance, as well as operate the property. Also, limitations are usually placed upon the operator's discretion to make certain decisions (e.g., to plug and abandon a well or release a portion of a lease). In addition, most operating agreements place a monetary limit on the authority of the operator (e.g., the operator may not undertake projects estimated to involve the expenditure of more than $10,000 without permission of the other owners, except in emergency).

b. Initial Drilling

Operating agreements are usually entered into only after the parties have decided to drill an initial well. Therefore, the operating agreement almost always directs the lessee to drill an initial well on the premises, setting time limits within which work must be commenced (to preserve the leases) and specifying depths to be reached and formations to be tested. Generally, consent of all of the parties is required to plug and abandon a well.

c. Additional Development

The possibility of development on the leased premises after the initial drilling is completed must be addressed. Also, most operating agreements cover substantial tracts of land with room for development drilling. Even if the operating agreement covers only a single drilling and spacing unit, there is always the possibility that it will be desirable in the future to drill again to a deeper depth.

D. DRILLING CONTRACTS

Drilling contracts are agreements for the drilling of a well or wells entered into by drilling contractors, who own drilling rigs and associated equipment, and persons or entities owning mineral or leasehold rights.

1.   KINDS OF DRILLING CONTRACTS

Drilling contracts are commonly structured to provide compensation on a daywork, footage, or turnkey basis. The provision for compensation generally controls the scope of discretion given the operator and affects the potential liability of the party contracting to have the well drilled.

One form of drilling contract is a *daywork contract*. Under a daywork drilling contract, the drilling contractor is compensated on the basis of the amount of time spent in drilling operations. In essence, the drilling party hires the drilling rig and its staff to work under his direction. Broad discretion is given to the contracting party to give instructions to the drilling contractor as to how to conduct the drilling operations. Broad liability follows.

Under a *footage contract*, the drilling contractor's compensation is calculated on the basis of the number of feet drilled. Under a footage contract, the party hiring the drilling contractor has less discretion than under a daywork contract to instruct the drilling contractor as to how drilling operations are to be conducted. Therefore, his liability is correspondingly narrower.

Under a *turnkey contract*, the drilling contractor agrees to drill, complete, equip, and deliver a well to the contracting party. The contracting party has little or no discretion to instruct the drilling contractor and little or no liability exposure for the contractor's actions.

2.　MODEL FORM DRILLING CONTRACTS

Two sets of model form drilling contracts are frequently encountered. One set of forms is prepared by the International Association of Drilling Contractors, the drilling industry association. Model forms are also prepared by the American Petroleum Institute, an association composed mainly of producers who must contract with drilling contractors. It is generally accepted that the form contracts are slanted in favor of the interests of the group that sponsored their preparation. Many drilling contractors and many producers have their own drilling contract forms, which are usually based upon one of the model forms.

The drilling industry is extremely competitive, and competition has great impact upon the terms of drilling contracts. In the late 1970's and through 1981 there was high demand for drilling rigs. Many drilling contractors refused to consider anything but daywork contracts, which effectively shifted most of the responsibility and risks to the producers. By the end of 1982, the drilling industry had experienced a severe recession, and nearly half of the country's drilling rigs were not working. In a period of a few months, drilling contractors' prices dropped 25 to 40 percent and producers found it possible to make substantial revisions to drilling contractors' agreements.

E. GAS CONTRACTS

A gas contract is an agreement for the sale of natural gas by a producer to a pipeline or end-user.

Gas contracts are among the most complicated oil and gas agreements. There are two reasons. First, because natural gas must be sold into pipelines that are expensive to construct and maintain, gas contracts are generally for a relatively long term, of five to fifteen years. Long term agreements are inherently complicated. Second, gas contracts are complicated by governmental regulation. Many of the terms and conditions of gas contracts are governed by state or federal regulatory requirements. As a result, draftsmen must try to comply with present and potential regulatory requirements as well as express clearly the agreement. Of course, neither the attempt to foresee all long-range problems nor the effort to comply with all present and future regulatory provisions can be completely successful. However, the length and complexity of gas contracts is a testament to the attempt.

1. COMMON SUBSTANTIVE ISSUES

Although there is wide variety in the terms found in gas contracts, there are common issues that will always be addressed in such agreements. They are (a) price, (b) take or pay obligations, (c) reserves committed, (d) reservations of seller, and (e) conditions of deliveries.

a. Price

The price to be paid to the producer for natural gas sold is of obvious importance both to the producer and to the pipeline purchaser. It is the major factor determining profitability for both. However, if the price of the natural gas in question is subject to regulatory controls, as most gas currently is, the producer can collect only the maximum regulated price permitted, even if the contract specifies a higher price.

(1) Price Escalation

Because gas contracts are usually long term, there is almost always provision for periodic adjustment of the initial price. Such provisions are referred to as *price escalation clauses.* There are several kinds of price escalation clauses, more than one of which may be included in the same contract. A *fixed escalation clause* provides for periodic increases in price of a stated amount or percentage; (e.g., 4¢ per quarter or 1½% per quarter). An *area rate clause* is one that provides for adjustment to the highest price that is permitted in the area by the appropriate regulatory body. A *most favored nations clause* provides for adjustment of the contract price upward if any other producer in the area receives a higher price for gas of similar quantity and quality. A *two-party most favored nations clause* adjusts the price only if a higher price is paid by the pipeline purchaser with whom the producer has contracted. A *third-party most favored nations clause* adjusts the price upward if any pipeline purchaser pays a high price. A *price redetermination clause* allows renegotiation of price on a periodic basis so that parties can take into account changed conditions in the market, and an *index-escalation clause* provides for adjustment of price in accordance with changes in an index or in some price (e.g., no. 2 fuel oil) the parties have agreed is an appropriate indicator of the value of natural gas.

(2) Deregulation

Since 1978, virtually all natural gas has been subject to price regulation at the wellhead. Under the provisions of the Natural Gas Policy Act of 1978, deregulation of certain categories of regulated natural gas is to begin December 31, 1984. Therefore, most current gas contracts contain provisions setting forth the agreement of the parties as to how the price of the natural gas is to be determined if deregulation takes place. Deregulation clauses usually contain one or more of the price escalation provisions discussed above.

(3) Buyer Protection

Because the price of natural gas had been held artificially low by price regulation under the Natural Gas Act of 1938, enactment of the Natural Gas Policy Act of 1978 brought burgeoning prices and increased drilling. Many pipelines (and many consumers) saw the prices they had to pay for natural gas increase sharply. Pipeline companies became concerned that they would commit themselves to prices too high to be economically feasible. They began inserting provisions which would permit adjustment of the price downward in that event.

There are two types of such clauses. A *FERC-out clause* provides that the pipeline need not pay any higher price than the appropriate regulatory agency will permit it to pass on to its customers. In other words, even though the price provided in the contract is less than the maximum regulated price, if a regulatory agency rules that that price is too high to be included in the pipeline's cost of service for rate making purposes, then it will be adjusted downward. A *market-out clause* permits the pipeline to redetermine the price downward if the price it pays is so high that the gas cannot be sold in the pipeline's main market area. Often, market-out provisions are keyed to some percentage of the price for competing fuels (e.g., 85% of no. 2 fuel oil) and permit the producer to terminate the contract and sell his gas elsewhere if he is displeased with the redetermination.

2.   TAKE OR PAY PROVISIONS

The second major issue addressed by most gas contracts is the obligation of the purchaser to take or to pay for gas that it does not purchase. The demand for natural gas is cyclical, dependent upon weather and economic conditions. Therefore, most producers seek to obtain promises from their pipeline purchasers that they will either *take* substantial amounts of gas or *pay* the producer for the amounts not taken. Take or pay provisions assure the producer of minimal cash flows. Usually, the purchaser has the right to make up quantities of natural gas paid for but not taken, with payment to the seller for the difference between the price in effect at the time the gas is taken and the price paid.

3. RESERVES COMMITTED

The reserves committed under the terms of a gas contract are particularly important to the pipeline purchaser because of its statutory obligation to continue service to its customers. Therefore, pipelines generally seek a broad commitment of reserves to the contract. A typical pipeline contract commits all gas that may be produced from present or future wells drilled to any formation on the property covered by the leases committed. Producers generally would prefer to commit only gas from the particular formation being produced by the particular well or wells that have been drilled.

4. RESERVATIONS

The issue of reservations is important to both the producer and pipeline for the same reason as is the reserves committed. The pipeline generally wishes to maximize the amount of reserves it controls for future deliveries. Therefore, it seeks to limit the purposes for which the producer may reserve and use gas. However, for the producer, it may be important to reserve the right to make available to the lessor "free gas" for farming or domestic purposes or to retain the right to use natural gas in oil lifting operations.

5.   CONDITIONS OF DELIVERY

Conditions of delivery refer to the quality of the gas and where it is to be delivered. Both issues are important economic factors. Removing impurities from the gas stream can be very expensive and increase the value of the gas substantially. So can transportation of the gas to a pipeline. Therefore, contract provisions for conditions of delivery may be the subject of hard bargaining.

F.   GAS BALANCING AGREEMENTS

1.   THE GAS BALANCING PROBLEM

Gas balancing agreements address the problem of the split-stream gas well. Often, cotenants separately dispose of their share of natural gas produced by joint operations. It is common for cotenants to sell their respective shares of natural gas to different pipeline companies upon different terms and for different prices. Such sales are called "split-stream" sales.

Although ownership of the gas stream may be split, natural gas is fungible, and it is impossible

[*361*]

to differentiate between gas owned by different cotenants. As a result, the gas owned by a co-tenant who does not wish to sell his share or whose purchaser is not in a position to take his share cannot be physically withheld; whoever buys gas produced takes gas that is owned not only by his seller but also other cotenants. As a result, it is inevitable that imbalances occur.

The custom of the oil and gas industry has been to provide *balancing in kind*. From time to time, the operator will adjust deliveries from one pipeline to another so that in the long run gas deliveries to various sellers are roughly in balance. However, occasionally disputes arise whether there is a legal right to balancing in kind. Or, what are the rights of the parties if there are insufficient reserves to bring deliveries into balance? Should there then be a *cash balancing?* If so, on what basis—on the basis of the amount received for the gas when sold (perhaps with an interest adjustment) or on the basis of value that the gas would have had had it been available when the balancing takes place? There are no clear-cut legal answers to these questions.

2. COMMON SUBSTANTIVE PROVISIONS

With increasing frequency, cotenants of wells producing natural gas enter into gas balancing agreements to provide for orderly periodic balancing. A typical agreement might require that the operator maintain an account showing amounts overproduced and underproduced. Generally, underproduced parties are given specific right to make up in kind, with cash balancing on the basis of amounts actually received where it is not possible to balance in kind. There are no "standard" provisions, however.

G. DIVISION ORDERS

A division order is a statement executed by all parties who claim an interest stipulating how proceeds of production are to be distributed. Division orders protect purchasers of production and those who distribute proceeds by warranting title to production transferred and indemnifying them for payments made. As is discussed in conjunction with the market value royalty problem at pages 272–274, division orders are sometimes used to attempt to amend oil and gas leases or ratify gas contracts.

It is not clear that division orders are true contracts. Oil division orders have been described as an offer for a unilateral contract that can be accepted by a purchaser who takes oil and sends payment for it. That description does not fit gas division orders as they affect royalty owners, however, for the lessor under an oil and gas lease does not generally have the right to take gas in-kind and so cannot sell it.

*

GLOSSARY OF OIL AND GAS TERMS

Acreage Contribution Agreement A support agreement by which the contributing party agrees to contribute leases or interests in leases in the area of a test well in exchange for information, if a well is drilled to an agreed depth. See also Support Agreements.

Ad Coelum Doctrine The common law doctrine that the owner of property owns from the heavens to the core of the earth and all of the elements within. Still accepted as the governing principle for "hard" minerals, but replaced by the rule of the capture for "fugacious" minerals such as oil and gas.

After-Acquired Title Clause An oil and gas lease clause that extends the coverage of the lease to any interest in the described property acquired after the lease. A common formulation is "This lease covers all the interest now owned by or hereafter vested in the lessor . . . ."

Apportionment Rule The rule followed in a minority of states (California, Mississippi and Pennsylvania) that royalties that accrue under a lease on property that has been subdivided after the grant are to be shared by the owners of the property proportionately to their interest in the property. For example, if O leases to A Company and subsequently sells the east half of the property to

X, O and X would share equally any future royalties from the property no matter where the wells were located. See also Non-Apportionment Rule.

Area Rate Clause　An indefinite price escalation clause found in some gas contracts that provides for increase of the contract price if any regulatory agency permits or prescribes a higher price for gas sold in the same area. Under the regulatory scheme of the Natural Gas Act, from 1961 to 1978, maximum prices were set by the Federal Power Commission (FPC) and its successor, the Federal Energy Regulatory Commission (FERC), on an area basis. Area rate clauses were drafted to permit sellers to collect the highest price permitted by the regulatory authority in the relevant area.

Assignment Clause　Another name for the Change of Ownership Clause.

Bonus　A payment to induce the lessor to execute the lease.

Bottom Hole Agreement　A support agreement in which the contributing party agrees to make a cash contribution in exchange for geological or drilling information, if a well is drilled to an agreed depth. See also Support Agreements.

Carried Interest　A fractional interest, usually in an oil and gas lease, free of some or all costs, which are borne by the remainder of the working interest. A common arrangement in drilling ventures is that the promoter will be "carried" to the casing point for $1/4$ of the working interest; i.e., the investors will pay 100% of the drilling costs for 75% of the working interest. The promoter will bear his 25% share of completion and operating costs under such an arrangement. The point to which the interest is to be carried is crucial to the economic effect of a carried interest. Also, there may be adverse tax impacts from the use of such interests.

Casing An industry term for pipe placed in a wellbore hole. *Surface casing* is to protect potable waters against pollution from drilling and producing operations. *Intermediate casing* protects deeper formations. *Production casing* is the pipe through which oil and gas is produced.

Casinghead Gas Gas produced from the casinghead (the top) of an oil well. Casinghead gas is natural gas held in solution with oil in the producing formation. At production or shortly after, it separates from oil.

Casing Point The point at which the well has been drilled to the desired depth and a decision must be made whether or not to place production pipe, called casing, in the hole and proceed to complete and equip the well for production.

Cessation of Production Clause An oil and gas lease clause that specifies what the lessee must do to maintain the lease in the event that production ceases. The purpose of the clause is to make more certain the temporary cessation of production doctrine. See also Temporary Cessation of Production Doctrine.

Change of Ownership Clause An oil and gas lease clause specifying what notice must be given by the lessor or his assignee to the lessee of changes in ownership in order to bind the lessee to recognize them. The purpose of the clause is to protect the lessee against the consequences of making an improper payment under the lease.

Continuous Operations Clause A form of operations clause.

Contribution Agreement Another name for a support agreement.

Correlative Rights Doctrine The corollary to the Rule of Capture that the right to capture oil and gas from potentially producing formations under one's property is subject to the concomitant duty to exercise the right without negligence or waste. See also Rule of Capture.

Cost Depletion Recovery of one's basis in a producing oil or gas well by deducting it proportionately over the producing life of the well. See also Percentage Depletion.

Cover All Clause Another name for a Mother Hubbard Clause.

Damages Clause A lease clause that imposes a duty on the lessee to pay the lessor or the surface owner for damage, usually specified, to the surface. Often such clauses are limited to "growing crops." In the absence of a damages clause the oil and gas lessee has no legal obligation to pay for "reasonable" damage to the surface necessary to obtain oil and gas; the lessee has an implied right to use the surface for oil and gas operations.

Daywork Drilling Contract A drilling contract under which the drilling contractor is compensated on the basis of the amount of time spent in drilling operations. Essentially, the lease owner hires the drilling rig and its staff to work under his direction. Broad discretion is given to the contracting party to give instructions to the drilling contractor as to how to conduct drilling operations. The courts impose broad liability upon the contracting party as a result of his broad discretion. See also Drilling Contracts.

Delay Rental A payment from the lessee to the lessor to maintain the lease from period to period during the primary term without drilling. See also Drilling-Delay Rental Clause, "Unless" lease and "Or" lease.

Deregulation Clause A gas contract provision spelling out how price is to be determined and what obligations the buyer and seller will owe one another in the event that regulated natural gas is freed from regulation.

Distillate The "wet" element of natural gas that may be removed as a liquid. Used interchangeably with "condensate" and "natural gasoline." Also, any product of the process of distillation.

Division Order An authorization to one who has a fund for distribution from persons entitled to the fund as to how it is to be distributed. In the oil and gas industry, division orders are entered into by royalty owners to sell oil and to give instructions for payment of delay rentals, royalties and other payments under a lease. Working interest owners also commonly sign division orders to give instructions to the purchaser of production for payment of the proceeds of sale.

Double Fraction Problem   A common ambiguity in conveyances that arises when one who owns a fractional interest conveys or reserves a fraction. For example, if one who owns an undivided $\frac{1}{2}$ interest in minerals conveys "an undivided $\frac{1}{2}$ interest in the minerals," there is an ambiguity whether the intention is to convey a half interest in all of the minerals or half of the grantor's half.

Drilling Contracts   Agreements for the drilling of a well or wells entered into between drilling contractors, who own drilling rigs and associated equipment, and persons or entities owning minerals or leasehold rights. The drilling contract spells out the rights and duties of the parties. Drilling contracts are generally structured to provide compensation on a daywork, footage, or turnkey basis. The provision for compensation generally controls the scope of discretion given to the operator and the amount of potential liability imposed on the party contracting to have the well drilled, as well as method of payment. See also "Daywork Drilling Contract," "Footage Drilling Contract" and "Turnkey Drilling Contract."

Drilling-Delay Rental Clause   A clause in an oil and gas lease that gives the lessee the right to maintain the lease from period to period during the primary term either by commencing drilling operations or by paying delay rentals. Drilling-delay rental clauses are inserted by lessees because the courts have said that they obviate any implied covenant to drill a test well on the premises. They are accepted by lessors because they provide for periodic rental payments. See also "Unless" lease and "Or" lease.

Dry Hole Agreement A support agreement in which the contributing party agrees to make a cash contribution in exchange for geological or drilling information, if a dry hole is drilled. See also Support Agreements.

Dry Hole Clause A provision in an oil and gas lease specifying what a lessee must do to maintain the lease for the remainder of the primary term after drilling an unproductive well. A dry hole clause is intended to make clear that the lease may be maintained by payment of delay rentals for the remainder of the primary term.

Duhig Rule A rule of title interpretation developed to deal with the frequent problem of over-conveyances of fractional interests. As stated in *Duhig v. Peavy-Moore Lumber Co., Inc.,* 135 Tex. 503, 144 S.W.2d 878 (1940), it may be summarized as follows:

> Where full effect cannot be given to the granted interest because of a previous outstanding interest, priority will be given to the granted interest (rather than to the reserved interest) until full effect is given to the granted interest.

The rule is not accepted in all states and is limited to conveyances by warranty deed.

Entirety Clause   A clause in an oil and gas lease or in a deed that states the agreement of the parties that royalties are to be apportioned.   The purpose of the clause from the viewpoint of the lessee is to make clear that the lessee's duties will not be increased by transfer by the lessor of a part of the leased premises.   From the viewpoint of the lessor, the purpose of the clause is to avoid the non-apportionment rule.

Escalation Clause   A gas contract provision providing for increase of the base price provided for in the agreement.   Escalation clauses are necessary because gas contracts are generally for long terms.

Farmout Agreement   An agreement by which one who owns an oil and gas lease (the farmor) agrees to assign to another (the farmee) an interest in the lease in return for drilling and testing operations on the lease.   From the viewpoint of the farmor, a farmout agreement may be used to maintain a lease by securing production or to comply with the implied covenant to develop or offset, or to obtain an interest in production without costs.   From the viewpoint of the farmee, a farmout agreement may be used to earn acreage not otherwise available or at lower cost than would otherwise be possible.   It may also serve as a device to keep people and equipment gainfully employed.

Fee Interest Literally, an interest in property of potentially infinite duration. As used in the oil and gas industry, it usually refers to ownership of both the surface interest and the mineral interest.

FERC The Federal Energy Regulatory Commission, the successor agency to the Federal Power Commission (FPC) and the agency responsible for administering the Natural Gas Act and the Natural Gas Policy Act.

FERC-Out Clause A gas contract clause that provides that the price paid to the producer shall be reduced (or the contract terminated) to the extent that the Federal Energy Regulatory Commission or other regulatory agencies will not permit it to be included in the gas pipeline's cost of service (and, in effect, passed on to consumers).

Footage Drilling Contract A drilling contract under which the drilling contractor is compensated on the basis of the footage drilled; e.g., $14 per foot. The contractor is hired to drill to a specified formation or depth and is given broad discretion to make the management decisions necessary to accomplish the taste. The risk of unexpected delays, as well as most liabilities, is upon the contractor rather than the contracting party.

Force Majeure Clause An oil and gas lease clause that provides that the lessee will not be held to have breached the lease terms while the lessee is prevented by *force majeure* (literally, "superior force") from performing. Typically, such clauses specifically indicate problems beyond the reasonable control of the lessee that will excuse performance.

FPC Clause   See Area Rate Clause.

Free Gas Clause   An oil and gas lease clause, found commonly in leases on properties in colder states, that entitles the lessor or the surface owner to use without charge gas produced from the leased property.   Such clauses are usually limited either as to the uses permitted (e.g. domestic heating and light) or as to the quantity that may be taken (e.g., not more than 300 MCF per year) or both.

Freestone Rider   Another name for a Pugh Clause.

Further Exploration Covenant   An implied oil and gas lease covenant that, once production has been obtained from the leased premises, the lessee will continue to explore other parts of the property and other formations under it.   In some jurisdictions the courts have said that the covenant for further exploration does not exist independently of the covenant for reasonable development.   It has been argued, however, that the covenants are separate and that the prudent operator standard does not apply to the covenant to further explore.   See also Reasonable Development Covenant and Prudent Operator Standard.

Gas Balancing Agreement   A contract among owners of the production of a gas well setting forth their agreement for the balancing of production in the event that one owner is able to sell more of the gas stream than other owners.

Gas Contract An agreement for the sale of natural gas by a producer to a pipeline or end-user.

Granting Clause The clause in the oil and gas lease that spells out what rights are given by the lessor to the lessee. Typically, an oil and gas lease granting clause will specify kinds of uses permitted and substances covered by the lease.

Habendum Clause Another name for the Term Clause.

Horsehead See Pumping Unit.

Intangible Drilling Costs Costs incurred incident to and necessary for the drilling and preparation of oil or gas wells for production that have no salvage value. By § 612 of the Internal Revenue code, such costs may be deducted in the year paid rather than capitalized and depreciated.

Landman A position in the oil and gas industry the responsibilities of which include acquiring oil and gas leases, negotiating arrangements for development of leases, and general management of leased properties.

Landowner's Royalty A share of oil or gas produced, free of costs of production, provided for the lessor in the royalty clause of the oil and gas lease. Traditionally, except in California, the landowner's royalty has been ⅛ of gross production for oil and ⅛ of the proceeds received from the sale of gas. Today, however, the size of the landowner's royalty is often negotiated. See also Royalty Interest.

Leasehold Interest Another name for the Working Interest.

Leasehold Royalty Another name for Landowners Royalty.

Lesser Interest Clause A lease clause that permits the lessee to reduce payments under the lease proportionately if the lessor has less than 100% of the mineral interest. Sometimes called a proportionate reduction clause.

Market-Out Clause A gas contract clause that provides that if the contract price for the gas purchased (plus certain costs incurred in getting it to the pipeline's principal market) exceeds an amount that will permit the gas to be sold, the contract price will be redetermined. Often such clauses are drafted by reference to competing fuels, e.g., fuel oil.

Marketing Covenant The promise implied in oil and gas leases that the lessee will market the production from the lease within a reasonable time and at a reasonable price. See also Prudent Operator Standard.

MCF The abbreviation for one thousand cubic feet, one of the standard units of measurement for natural gas.

Mineral Acre The full mineral interest in one acre of land.

Mineral Interest   The right to search for, develop and produce oil and gas (and other minerals) from the land, as well as (in some states) the right to present possession of the oil and gas in place.   It is the mineral interest that is granted by the oil and gas lease.   See also Fee Interest and Surface Interest.

Mineral Servitude   Under the Louisiana Mineral Code, a charge upon land in favor of a person or another tract of land that creates a limited right to use of the land to explore for and produce minerals.   Generally equivalent to a severed mineral interest in a common law state.

Mother Hubbard Clause   A lease clause to protect the lessee against errors in description of property by providing that the lease covers all the land owned by the lessor in the area.   Sometimes called a "cover all" clause.   Sometimes combined with an after-acquired title clause.

Net Profits Interest   A share of production free of the costs of production expressed as a fraction or a percentage of production, to the extent that there is a net profit.   The methodology of defining "net profits" is crucial to a net profits interest.

Non-Apportionment Rule   The rule followed in the majority of states that royalties that accrue under a lease on property that has been subdivided after the grant of a lease are *not* to be shared by the owners of the various subdivisions but belong exclusively to the owner of the subdivision upon which the well is located that produces the royalty. For example, if O leases to A Company and subsequently sells the east half of the property to X, X would take all royalties that accrued to a well drilled on the east half and O would take all royalties that accrued to a well drilled on the west half. The non-apportionment rule may be modified by an entirety clause. See also Apportionment Rule.

Non-Ownership Theory The characterization of oil and gas rights that a severed mineral interest owner has merely a right to search, develop and produce oil and gas from the land, but not a present right to possess the oil and gas in place. Because there is no right to present possession, the interest of a severed mineral interest owner in a non-ownership theory state is akin to a *profit a prendre*, a right to use the land and remove items of value from it. Adopted in California, Louisiana and Oklahoma, as well as other producing states.

Non-Participating Royalty A share of production, free of the costs of production, carved out of the mineral interest. A non-participating royalty holder is entitled to the stated share of production without regard to the terms of any lease. Such royalties are often retained by mineral interest owners who sell their rights. See also Royalty and Overriding Royalty.

Obstruction A common law doctrine that suspends the running of time under a lease or extends the lease for a reasonable period of time where rights granted under a lease are interfered with by the lessor or one claiming through the lessor.

Operating Agreement A contract among owners of working interest in a producing oil or gas well setting forth the parties' agreement as to drilling, development, operations and accounting.

Operations Clause A clause frequently found in oil and gas leases that provides that the lease will be continued so long as operations for oil and gas development continue on the premises. There are numerous variations. Two common ones are the well completion clause and the continuous operations clause. The former provides that a lessee who commences drilling prior to termination of the lease has the right to complete the well and to maintain the lease if the drilling achieves production. The latter gives the lessee the right not only to continue drilling a well begun prior to termination but to commence additional wells too.

"Or" Lease An oil and gas lease with a drilling-delay rental clause structured so that the lessee promises to commence drilling operations *or* to pay delay rentals from time to time during the primary term. If the lessee fails to do one or the other, the lease does not automatically terminate; instead the lessee is liable to pay the delay rental amount.

Overriding Royalty A share of production, free of the costs of production, carved out of the lessee's interest under an oil and gas lease. Overriding royalty interests are frequently used to compensate landmen, lawyers, geologists or others who have helped to structure a drilling venture. An overriding royalty interest terminates when the underlying lease terminates. See also Royalty and Non-participating Royalty.

Ownership in Place Theory The characterization of oil and gas rights that the owner of severed rights owns the right to present possession of the oil and gas in place as well as the right to use the land surface to search, develop and produce from the property. Adopted in Texas, New Mexico, Kansas, Mississippi, and other major producing states. Since oil and gas rights are subject to the rule of capture even in ownership in place theory states, the rights of a severed mineral interest owner to oil and gas in such states are often described as a fee simple determinable.

Paid-Up Lease An oil and gas lease that does not contain provision for payment of delay rentals. The lease is effective for the whole period of the primary term without payments of delay rentals.

Percentage Depletion A provision of § 611 of the Internal Revenue Code that permits a taxpayer who owns an economic interest in a producing oil or gas well to deduct a specified percentage of the gross income from the well in lieu of depleting his actual basis. See also Cost Depletion.

Petroleum Conservation Laws State laws that limit the rule of capture and define the correlative rights doctrine by regulating the drilling and operation of oil and gas wells.

Pooling Bringing together, either by voluntary agreement (voluntary pooling) or by order of an administrative agency (compulsory or forced pooling), small tracts or fractional interests to drill a well. Pooling is usually undertaken to comply with well spacing requirements established by state law or regulation. Pooling is usually associated with drilling a single well and operating that well by primary production techniques. In the oil and gas industry, the term is often used interchangeably with unitization.

Pooling Clause A clause found in most leases that grants a power of attorney to the lessee to combine part or all of the leased acreage with other properties for development or operation.

Prescription A doctrine of Louisiana law that extinguishes unused mineral servitudes after 10 years. To interrupt the running of the prescription period, operations to discover or produce on the land or land pooled with it are required.

Price Redetermination Clause A clause in a gas contract that provides for price redetermination from time to time or upon election of one of the parties.

Primary Term The option period set by the oil and gas lease term clause during which the lessee retains the right to search, develop and produce from the premises without having any obligation to do so. The primary term should be sufficiently long to permit the lessee to evaluate the property and make arrangements to drill it. In practice, the primary term may extend for 24 hours or

25 years, depending upon how much competition there is for leases in the area. See also Term Clause and Secondary Term.

Production Payment A share of production from property, free of the costs of production, that terminates when an agreed sum has been paid; e.g., " $\frac{1}{5}$ of all oil and gas produced and saved from said land until the market value at the well of such production shall aggregate One Million Dollars . . . ."

Proportionate Reduction Clause Another name for the Lesser Interest Clause.

Protection Covenant The promise implied in oil and gas leases that the lessee will protect the premises against drainage by drilling a producing well to the reservoir which is subject to drainage, if the reasonable prudent operator would do so. See also Prudent Operator Standard.

Prudent Operator Standard The test generally applied to determine a lessee's compliance with implied lease covenants. The standard is a reference to what the reasonable, competent operator in the oil and gas industry, acting in good faith and with economic motivation, would do under the circumstances.

Pugh Clause A lease clause (called a Freestone Rider in Texas) that modifies the effect of most lease pooling clauses by severing pooled portions of the lease from unpooled portions of the lease so that drilling or production on a pooled portion will not maintain the lease as to unpooled portions.

Pumping Unit Equipment used to pump oil to the surface when a pressure differential between the pressure in the formation and that in the borehole is insufficient to cause oil to rise up the borehole to the surface. Sometimes called a "pumpjack" or "horsehead."

Pumpjack Another term for a pumping unit.

Reasonable Development Covenant The promise implied in oil and gas leases that, once production is obtained, the lessee will continue to develop the premises as would the reasonable prudent operator rather than merely holding the lease by the production already obtained. See also Further Exploration Covenant and Prudent Operator Standard.

Royalty Interest A share of production free of the costs of production, when and if there is production; usually expressed as a fraction; e.g. $\frac{1}{6}$. In most states, however, a royalty is subject to costs for severance taxes, transportation, cleaning and compression (costs subsequent to production) unless the lease provides otherwise. A royalty interest has no right to operate the property. Therefore, a royalty has no right to lease or to share in bonus or delay rental. In some states a royalty interest has the right of access and egress to take the royalty production. There are several different kinds of royalty interests commonly encountered. See e.g. Landowner's Royalty, Non-participating Royalty, and Overriding Royalty.

Rule of Capture The fundamental principle of oil and gas law that there is no liability for drainage of oil and gas from under the lands of another, so long as there has been no trespass and all relevant statutes and regulations have been observed.

Secondary Term The term of the oil and gas lease after production has been established, typically "as long thereafter as oil and gas is produced from the premises." See also Term Clause and Primary Term.

Separator Equipment used at the well site to separate oil, water and gas produced in solution with oil. Simple separators simply heat oil to speed the natural separation process. More complex separators may use chemicals.

Severance A term used to describe a transfer or reservation of a part of the bundle of rights that make up property ownership from the whole. Mineral rights are frequently "severed" from property that may contain oil and gas or other minerals.

Shut-in Royalty Clause A lease clause that permits the lessee to maintain the lease while there is no production from the premises because wells capable of production are shut-in by making a payment of "shut-in royalty" in lieu of production.

Subrogation Clause A lease provision that permits the lessee to pay taxes, mortgages or other encumbrances on the leased property and to recover such payments out of future proceeds under the lease.

Support Agreements Contracts between people or entities in the oil and gas industry that encourage and "support" exploratory operations. Generally, one party agrees to contribute money or property to another if the other will drill a well on leases that he holds and provide the contributing party with information from tests conducted. Thus, from the viewpoint of the contributing party, a support agreement is a purchase of geological or technological information. From the viewpoint of the party receiving the support, the contribution lessens the cost or the risk of drilling operations. For further discussion see "Dry Hole Agreement," "Bottom Hole Agreement," and "Acreage Contribution Agreement."

Surface Interest All rights to property other than the mineral interest. The surface interest has the right to the surface subject to the right of the mineral interest owner to use the surface to search, develop and produce minerals. The surface interest is entitled to all substances found in or under the soil that are not defined as "minerals."

Surrender Clause A clause commonly found in an oil and gas lease authorizing the lessee to release his rights to all or any portion of the leased premises at any time and be relieved of further obligations as to the acreage surrendered.

Temporary Cessation of Production Doctrine The rule that an oil and gas lease term "for so long thereafter as oil and gas are produced" will not terminate once production is attained unless the cessation of production is for an "unreasonable" length of time, taking into account all of the facts and circumstances. See also Cessation of Production Clause.

Term Clause The clause of an oil and gas lease that defines for how long the interest granted to the lessee will extend. Modern oil and gas leases typically provide for a primary term, a fixed number of years during which the lessee has no obligation to develop the premises, and a secondary term for "so long thereafter as oil and gas produced," once development takes place.

Term Royalty A share of production free of the costs of production that is not perpetual. A term royalty may be for a fixed term (e.g., for 25 years) or defeasible (e.g., for 25 years and so long thereafter as there is production from the premises). See also Royalty Interest.

Top Lease A lease granted on property already subject to an oil and gas lease. Generally, a top lease grants rights if and when the existing lease expires.

Turnkey Drilling Contract A drilling contract under which the drilling contractor agrees to perform stated functions for an agreed price. The contracting party has little or no discretion to instruct the drilling contractor and little or no liability exposure for the contractor's actions.

Unitization Bringing together, either by voluntary agreement (voluntary unitization) or by order of an administrative agency (compulsory or forced unitization) some or all of the well spacing units over a producing reservoir for joint operations. Unitization is usually undertaken after primary production has begun to fall off substantially. It is usually undertaken to permit efficient secondary recovery operations. In the oil and gas industry, the term is often used interchangeably with Pooling.

Unitization Clause A clause in an oil and gas lease granting the lessee the right to unitize the leased premises. It is somewhat unusual to see "true" unitization clauses; generally the right to unitize is addressed tangentially in the pooling clause and is subject to an acreage limtation that makes it difficult to conduct secondary recovery operations.

"Unless" Lease An oil and gas lease with a drilling-delay rental clause structured as a special limitation on the primary term; unless delay rentals are paid or drilling operations are commenced from time to time as specified, the unless lease automatically terminates, though the lessee has no liability for his failure.

Warranty Clause A clause in an oil and gas lease by which the lessor guarantees that his title is without defect and agrees to defend it. If the warranty is breached, the lessor may be held liable to the lessee to the extent that the lessor has received payments under the lease. Presence of a warranty in an oil and gas lease may also cause after-acquired interest to pass from the lessor to the lessee by application of estoppel by deed.

Well Completion Clause A form of operations clause.

Working Interest The rights to the mineral interest granted by an oil and gas lease, so-called because the lessee acquires the right to work on the leased property to search, develop and produce oil and gas (and the obligation to pay all costs).

APPENDIX OF FORMS

1. Mineral Deed.
2. Conveyance of Nonparticipating Royalty Interest.
3. Oil and Gas Lease (Texas AAPL Form 675).
4. Oil and Gas Lease (Colorado AAPL Form 681).

MINERAL DEED

(APPROVED BY MID–CONTINENT ROYALTY OWNER'S ASSOCIATION)

KNOW ALL MEN BY THESE PRESENTS:

THAT _____

_____

_____

_____

of _____

(Give exact postoffice address)

hereinafter called Grantor, (whether one or more) for and in consideration of the sum of _____ Dollars, ($_____) cash in hand paid and other good and valuable considerations, the receipt of which is hereby acknowledged, do_____, hereby grant, bargain, sell, convey, transfer, assign and deliver unto _____ of _____,

(Give exact postoffice address)

hereinafter called Grantee, (whether one or more) an undivided _____ interest in and to all of the oil, gas and other minerals in and under and that may be produced from the following described lands situated in _____ County, State of _____, to-wit:

containing _____ acres, more or less, together with the right of ingress and egress at all times for the purpose of mining, drilling, exploring, operating and developing said lands for oil, gas, and other minerals, and storing, handling, transporting and marketing the same therefrom with the right to remove from said land all of Grantee's property and improvements.

This sale is made subject to any rights now existing to any lessee or assigns under any valid and subsisting oil and gas lease of record heretofore executed; it being understood and agreed that said Grantee shall have, receive, and enjoy the herein granted undivided interest in and to all bonuses, rents, royalties and other benefits which may accrue under the terms of said lease insofar as it covers the above described land from and after the date hereof, precisely as if the Grantee herein had been at the date of the making of said lease the owner of a similar undivided interest in and to the lands described and Grantee one of the lessors therein.

Grantor agrees to execute such further assurances as may be requisite for the full and complete enjoyment of the rights herein granted and likewise agrees that Grantee herein shall have the right at any time to redeem for said Grantor by payment, any mortgage, taxes, or other liens on the above described land, upon default in payment by the Grantor, and be subrogated to the rights of the holder thereof.

TO HAVE AND TO HOLD The above described property and easement with all and singular the rights, privileges, and appurtenances thereunto or in any wise belonging to said Grantee herein, _____ heirs, successors, personal representatives, administrators, executors, and assigns forever, and Grantor does hereby warrant said title to Grantee _____ heirs, executors, administrators, personal representatives, successors and assigns forever, and does hereby agree to defend all and singular the said property unto the said Grantee herein _____ heirs, successors, executors, personal representatives, and assigns against all and every person or persons whomsoever lawfully claiming or to claim the same, or any part thereof.

WITNESS Grantor's hand this _____ day of _____, 19__.

_____   _____

_____   _____

_____   _____

_____   _____

(ACKNOWLEDGMENT)

CONVEYANCE OF NONPARTICIPATING ROYALTY INTEREST

KNOW ALL MEN BY THESE PRESENTS:

That _____, hereinafter called "Grantors," for good and valuable consideration in hand paid, the receipt and sufficiency of which Grantors hereby acknowledge, do hereby grant, bargain, sell, convey, transfer, assign, and deliver unto _____, hereinafter called "Grantee," for the term hereinafter specified, an undivided _____ interest in any and all Royalty (as hereinafter defined) on oil, gas, casinghead gas, distillate, condensate, and any and all other hydrocarbon or nonhydrocarbon substances, whether similar or dissimilar which may be produced or extracted and saved from the following described land situated in _____ County, Oklahoma:

(Description)

or from lands pooled or unitized with any portion thereof, or from lands located within any governmental drilling and spacing unit which includes any portion thereof, together with the right of ingress and egress to the surface thereof for the purpose of taking and receiving the herein granted interest in production.

"Royalty," as used herein, shall mean: (1) Any interest in production or the proceeds therefrom reserved by or granted to Grantors, their successors or assigns, in connection with any present or future lease for the production or extraction of substances from said lands, including, but not limited to, whether royalty, net profits interest, or production payment; (2) any payment so granted or reserved to be paid in lieu of production, including, but not limited to, whether similar or dissimilar, any shut-in well payments or minimum royalty; (3) in the event of the development of any portion of the above described land by Grantors, their successors or assigns, for the production or extraction of any substances, the same interest in production, proceeds, or payment to which Grantee would have been entitled if

APPENDIX

Grantors, as of the date of commencement of development, had executed a lease providing for a royalty equivalent to that set forth in the succeeding paragraph. As to production or extraction of substances from lands pooled or unitized with the above described land or from lands located within any governmental drilling and spacing unit which includes any portion thereof, "Royalty" shall include only that portion of said production, proceeds or payments attributable to the above described land's interest in said production unit. "Royalty" shall not include any cash bonus received by Grantor at the time of executing any future oil, gas, or mineral lease, nor any rental paid for the privilege of deferring commencement of development under any existing or future lease.

Grantors, their successors and assigns, reserve the exclusive right to execute leases for the production or extraction of substances from the above described land; provided that no lease or contract for the development of said land shall provide for a royalty less than that customarily then being received by lessors in the area, and in no event less than one-eighth ($\frac{1}{8}$) of all substances produced and extracted, delivered free and clear of all cost and expense except a proportionate part of taxes on production; and provided further that Grantors, their successors and assigns in exercising said leasing power, shall be deemed to owe a fiduciary duty to Grantee.

TO HAVE AND TO HOLD unto Grantee, his successors and assigns, (forever) (for a term of—years from the date hereof, and as long thereafter as oil, gas or other minerals is being produced from, or a shut-in well is located on, or operations are being conducted on, the above described land, or from lands pooled or unitized with any portion thereof, or from lands located within any governmental drilling and spacing unit which includes any portion thereof); and Grantors do hereby warrant title to the herein granted interest to Grantee, his successors and assigns, and do hereby agree to defend all and singular such interest unto Grantee, his successors or assigns, against any person whomsoever claiming or to claim the same or any part thereof; and Grantors, on behalf of themselves, their successors and assigns, do hereby agree to execute such further assurances as may be requisite for the full and complete enjoyment of the herein granted interest.

APPENDIX

EXECUTED to be effective as of this _____ day of _____,
19__.

_____   _____

_____   _____

                                         _____

(ACKNOWLEDGMENT)

AAPL FORM 675

OIL AND GAS LEASE *

TEXAS FORM—SHUT-IN CLAUSE, POOLING CLAUSE

THIS AGREEMENT made and entered into the _____ day of _____, 19__, by and between _____, Lessor and _____, Lessee.

WITNESSETH:

1. Lessor, in consideration of the sum of _____ Dollars ($_____), in hand paid, receipt of which is hereby acknowledged, and the royalties herein provided, does hereby grant, lease and let unto Lessee for the purpose of exploring, prospecting, drilling and mining for and producing oil and gas and all other hydrocarbons, laying pipe lines, building roads, tanks, power stations, telephone lines and other structures thereon to produce, save, take care of, treat, transport and own said products, and housing its employees, and without additional consideration, does hereby authorize Lessee to enter upon the lands covered hereby to accomplish said purposes, the following described land in _____ County, Texas, to-wit:

[Legal Description]

This Lease also covers and includes any and all lands owned or claimed by the Lessor adjacent or contiguous to the land described hereinabove, whether the same be in said survey or surveys or in adjacent surveys, although not included within the boundaries of the land described above. For the purpose of calculating rental payments hereinafter provided for the lands covered hereby are estimat-

* Approved for use in Texas by the American Association of Petroleum Landmen.

ed to comprise _____ acres, whether it actually comprises more or less.

2. Subject to the other provisions herein contained this Lease shall be for a term of _____ years from this date (called "primary term") and as long thereafter as oil and gas or other hydrocarbons are being produced from said land or land with which said land is pooled hereunder.

3. The royalties to be paid by Lessee are as follows: On oil, one-eighth of that produced and saved from said land, the same to be delivered at the wells or to the credit of Lessor into the pipe line to which the wells may be connected. Lessee shall have the option to purchase any royalty oil in its possession, paying the market price therefor prevailing for the field where produced on the date of purchase. On gas, including casinghead gas, condensate or other gaseous substances, produced from said land and sold or used off the premises or for the extraction of gasoline or other products therefrom, the market value at the well of one-eighth of the gas so sold or used, provided that on gas sold at the wells the royalty shall be one-eighth of the amount realized from such sale. While there is a gas well on this Lease, or on acreage pooled therewith, but gas is not being sold or used Lessee shall pay or tender annually at the end of each yearly period during which such gas is not sold or used, as royalty, an amount equal to the delay rental provided for in paragraph 5 hereof, and while said royalty is so paid or tendered this Lease shall be held as a producing Lease under paragraph 2 hereof. Lessee shall have free use of oil, gas and water from said land, except water from Lessor's wells, for all operations hereunder, and the royalty on oil and gas shall be computed after deducting any so used.

4. Lessee, at its option, is hereby given the right and power to voluntarily pool or combine the acreage covered by this Lease, or any portion thereof, as to the oil and gas, or either of them, with other land, lease or leases in the immediate vicinity thereof to the extent hereinafter stipulated, when in Lessee's judgment it is necessary or advisable to do so in order to properly develop and operate said leased premises in compliance with the Spacing Rules of the Railroad Commission of Texas, or other lawful authorities, or when to do so would, in the judgment of Lessee, promote the conservation of oil and gas from said premises. Units pooled for oil hereunder shall not substantially exceed 80 acres each in area, and units pooled

for gas hereunder shall not substantially exceed 640 acres each in area plus a tolerance of ten per-cent thereof in the case of either an oil unit or a gas unit, provided that should governmental authority having jurisdiction prescribe or permit the creation of units larger than those specified, units thereafter created may conform substantially in size with those prescribed by governmental regulations. Lessee under the provisions hereof may pool or combine acreage covered by this Lease, or any portion thereof as above provided for as to oil in any one or more strata and as to gas in any one or more strata. The units formed by pooling as to any stratum or strata need not conform in size or area with the unit or units into which the Lease is pooled or combined as to any other stratum or strata, and oil units need not conform as to area with gas units. The pooling in one or more instances shall not exhaust the rights of Lessee hereunder to pool this Lease, or portions thereof, into other units. Lessee shall file for record in the county records of the county in which the lands are located an instrument identifying and describing the pooled acreage. Lessee may at its election exercise its pooling operation after commencing operations for, or completing an oil or gas well on the leased premises, and the pooled unit may include, but is not required to include, land or leases upon which a well capable of producing oil or gas in paying quantities has theretofore been completed, or upon which operations for drilling of a well for oil or gas have theretofore been commenced. Operations for drilling on or production of oil or gas from any part of the pooled unit composed in whole or in part of the land covered by this Lease, regardless of whether such operations for drilling were commenced or such production was secured before or after the execution of this instrument or the instrument designating the pooled unit, shall be considered as operations for drilling on or production of oil or gas from the land covered by this Lease whether or not the well or wells are actually located on the premises covered by this Lease, and the entire acreage constituting such unit or units, as to oil and gas or either of them as herein provided, shall be treated for all purposes except the payment of royalties on production from the pooled unit as if the same were included in this Lease. For the purpose of computing the royalties to which owners of royalties and payments out of production and each of them shall be entitled upon production of oil and gas, or either of them from the pooled unit, there shall be allocated to the land covered by this Lease and included in said unit a pro rata portion of the oil and gas, or either of them, produced from the pooled unit after deducting that used for operations on the pooled unit. Such allocation shall be on an acreage basis, that is to say, there shall be allocated to the acreage covered by this Lease and included in the pooled unit that pro rata portion of the oil and gas, or either of them, produced from the pooled unit which the number of surface acres covered by this Lease and included in the pooled unit bears to the total number of surface acres included in the pooled unit. Royalties hereunder shall be computed on the portion of such production, whether it be oil or gas or either of them, so allocated to the land covered by this Lease and included in the unit just as though such production were from such land. The production from an oil well will be considered as production from the Lease or oil pooled unit from which it is producing and not as production from a gas pooled unit; and production from a gas well will be considered as production from the Lease or gas pooled unit from which it is producing and not from the oil pooled unit.

5. If operations for drilling are not commenced on said land, or on acreage pooled therewith as above provided for, on or before one year from the date hereof, the Lease shall terminate as to both parties, unless on or before such anniversary date Lessee shall pay or tender to Lessor, or to the credit of Lessor in the _____ Bank at _____, Texas, (which Bank and its successors shall be Lessor's agent and shall continue as the depository for all rentals payable hereunder regardless of changes in ownership of said land or the rentals) the sum of _____ Dollars ($_____), herein called rentals, which shall cover the privilege of deferring commencement of drilling operations for a period of twelve (12) months. In like manner and upon like payment or tenders annually the commencement of drilling operations may be further deferred for successive periods of twelve (12) months each during the primary term hereof. The payment or tender of rental under this paragraph and of royalty under paragraph 3 on any gas well from which gas is not being sold or used may be made by check or draft of Lessee mailed or delivered to Lessor, or to said Bank on or before the date of payment. If such Bank, or any successor Bank, should fail, liquidate or be succeeded by another Bank, or for any reason fail or refuse to accept rental, Lessee shall not be held in default for failure to make such payment or tender of rental until thirty (30) days after Lessor shall deliver to

APPENDIX

Lessee a proper recordable instrument, naming another Bank as Agent to receive such payments or tenders. Cash payment for this Lease is consideration for this Lease according to its terms and shall not be allocated as a mere rental for a period. Lessee may at any time or times execute and deliver to Lessor, or to the depository above named, or place of record a release covering any portion or portions of the above described premises and thereby surrender this Lease as to such portion or portions and be relieved of all obligations as to the acreage surrendered, and thereafter the rentals payable hereunder shall be reduced in the proportion that the acreage covered hereby is reduced by said release or releases.

6. If prior to discovery of oil, gas or other hydrocarbons on this land, or on acreage pooled therewith, Lessee should drill a dry hole or holes thereon, or if after the discovery of oil, gas or other hydrocarbons, the production thereof should cease from any cause, this Lease shall not terminate if Lessee commences additional drilling or re-working operations within sixty (60) days thereafter, or if it be within the primary term, commences or resumes the payment or tender of rentals or commences operations for drilling or re-working on or before the rental paying date next ensuing after the expiration of sixty (60) days from the date of completion of the dry hole, or cessation of production. If at any time subsequent to sixty (60) days prior to the beginning of the last year of the primary term, and prior to the discovery of oil, gas or other hydrocarbons on said land, or on acreage pooled therewith, Lessee should drill a dry hole thereon, no rental payment or operations are necessary in order to keep the Lease in force during the remainder of the primary term. If at the expiration of the primary term, oil, gas or other hydrocarbons are not being produced on said land, or on acreage pooled therewith, but Lessee is then engaged in drilling or re-working operations thereon, or shall have completed a dry hole thereon within sixty (60) days prior to the end of the primary term, the Lease shall remain in force so long as operations are prosecuted with no cessation of more than sixty (60) consecutive days, and if they result in the production of oil, gas or other hydrocarbons, so long thereafter as oil, gas or other hydrocarbons are produced from said land, or acreage pooled therewith. In the event a well or wells producing oil or gas in paying quantities shall be brought in on adjacent land and draining the leased premises, or acreage pooled therewith, Lessee agrees to drill

such offset wells as a reasonably prudent operator would drill under the same or similar circumstances.

7. Lessee shall have the right at any time during or after the expiration of this Lease to remove all property and fixtures placed on the premises by Lessee, including the right to draw and remove all casing. When required by the Lessor, Lessee shall bury all pipe lines below ordinary plow depth, and no well shall be drilled within two hundred (200) feet of any residence or barn located on said land as of the date of this Lease without Lessor's consent.

8. The rights of each party hereunder may be assigned in whole or in part, and the provisions hereof shall extend to their heirs, successors and assigns, but no change or division in the ownership of the land, rentals or royalties, however accomplished, shall operate to enlarge the obligations, or diminish the rights of Lessee; and no change or division in such ownership shall be binding on Lessee until thirty (30) days after Lessee shall have been furnished with a certified copy of recorded instrument or instruments evidencing such change of ownership. In the event of assignment hereof in whole or in part, liability for breach of any obligation issued hereunder shall rest exclusively upon the owner of this Lease, or portion thereof, who commits such breach. In the event of the death of any person entitled to rentals hereunder, Lessee may pay or tender such rentals to the credit of the deceased, or the estate of the deceased, until such time as Lessee has been furnished with the proper evidence of the appointment and qualification of an executor or an administrator of the estate, or if there be none, then until Lessee is furnished satisfactory evidence as to the heirs or devisees of the deceased, and that all debts of the estate have been paid. If at any time two or more persons become entitled to participate in the rental payable hereunder, Lessee may pay or tender such rental jointly to such persons, or to their joint credit in the depository named herein; or, at the Lessee's election, the portion or part of said rental to which each participant is entitled may be paid or tendered to him separately or to his separate credit in said depository; and payment or tender to any participant of his portion of the rentals hereunder shall maintain this Lease as to such participant. In the event of an assignment of this Lease as to a segregated portion of said land, the rentals payable hereunder shall be apportioned as between the several leasehold owners ratably according to the surface area of each, and default in rental payment by one

APPENDIX

shall not affect the rights of other leasehold owners hereunder. If six or more parties become entitled to royalty payments hereunder, Lessee may withhold payment thereof unless and until furnished with a recordable instrument executed by all such parties designating an agent to receive payment for all.

9. The breach by Lessee of any obligations arising hereunder shall not work a forfeiture or termination of this Lease nor cause a termination or reversion of the estate created hereby nor be grounds for cancellation hereof in whole or in part unless Lessor shall notify Lessee in writing of the facts relied upon in claiming a breach hereof, and Lessee, if in default, shall have sixty (60) days after receipt of such notice in which to commence the compliance with the obligations imposed by virtue of this instrument, and if Lessee shall fail to do so then Lessor shall have grounds for action in a court of law or such remedy to which he may feel entitled. After the discovery of oil, gas or other hydrocarbons in paying quantities on the lands covered by this Lease, or pooled therewith, Lessee shall reasonably develop the acreage retained hereunder, but in discharging this obligation Lessee shall not be required to drill more than one well per eighty (80) acres of area retained hereunder and capable of producing oil in paying quantities, and one well per six hundred forty (640) acres of the area retained hereunder and capable of producing gas or other hydrocarbons in paying quantities, plus a tolerance of ten percent in the case of either an oil well or a gas well.

10. Lessor hereby warrants and agrees to defend the title to said lands and agrees also that Lessee at its option may discharge any tax, mortgage or other liens upon said land either in whole or in part, and in the event Lessee does so, it shall be subrogated to such lien with the right to enforce same and apply rentals and royalties accruing hereunder towards satisfying same. Without impairment of Lessee's rights under the warranty in event of failure of title, it is agreed that if Lessor owns an interest in the oil, gas or other hydrocarbons in or under said land, less than the entire fee simple estate, then the royalties and rentals to be paid Lessor shall be reduced proportionately. Failure of Lessee to reduce such rental paid hereunder or over-payment of such rental hereunder shall not impair the right of Lessee to reduce royalties payable hereunder.

11. Should Lessee be prevented from complying with any express or implied covenant of this Lease, from conducting drilling, or re-

working operations thereon or from producing oil or gas or other hydrocarbons therefrom by reason of scarcity of, or inability to obtain or to use equipment or material, or by operation of force majeure, or because of any federal or state law or any order, rule or regulation of a governmental authority, then while so prevented, Lessee's obligations to comply with such covenant shall be suspended, and Lessee shall not be liable in damages for failure to comply therewith; and this Lease shall be extended while and so long as Lessee is prevented by any such cause from conducting drilling or reworking operations on, or from producing oil or gas or other hydrocarbons from the leased premises; and the time while Lessee is so prevented shall not be counted against the Lessee, anything in this Lease to the contrary notwithstanding.

IN WITNESS WHEREOF this instrument is executed on the date first above set out.

_____   _____

_____   _____

_____

(ACKNOWLEDGMENT)

AAPL FORM 681
OIL AND GAS LEASE *
COLORADO—SHUT–IN ROYALTY, POOLING

THIS AGREEMENT made and entered into this _____ day of
_____, 19__ by and between _____,
hereinafter called Lessor (whether one or more) and _____
hereinafter called Lessee (whether one or more).

WITNESSETH:

1. Lessor, for and in consideration of the sum of _____ Dollars
($_____) in hand paid, the royalties provided herein, and the cove-
nants of the Lessee, hereby grants, leases and lets exclusively to Lessee the land described below for the purpose of investigating, exploring for, drilling for, producing, saving, owning, handling, storing,
treating and transporting Oil and Gas together with all rights, privileges and easements useful for Lessee's operations on said land and
on land in the same field with a common Oil and Gas Reservoir,
including but not limited to rights to lay pipelines, build roads, construct tanks, pump and power stations, power and communication
lines, houses for its employees, and other structures and facilities, and
the right to drill for, produce and use fresh water. The Phrase "Oil
and Gas" as used herein includes all hydrocarbons and other substances produced therewith. The land included in this Lease is situated in _____ County, State of Colorado, and is described as follows:

[Description]

including all Oil and Gas and substances produced therewith underlying lakes and streams of which all or any part of the land is ripari-

* Approved for use in Colorado by the American Association of Petroleum Landmen.

APPENDIX

an, all roads, easements, and rights-of-way which traverse or adjoin said land and including all lands owned or claimed by Lessor as a part of any of said land, and including all reversionary rights therein, said land containing _____ acres more or less. This lease covers all the Interest now owned by, or hereafter vested in the Lessor and Lessor releases and waives all rights under any Homestead Exemption Laws. In calculating any payments based on acreage, Lessee may consider that the land contains the acreage stated above, whether it actually contains more or less. Lessee may inject water, salt water, gas or other substances into any stratum or strata under said land and not productive of fresh water.

2. This Lease shall remain in force for a period of ten (10) years from this date, called "primary term", and as long thereafter as Oil, Gas or other Hydrocarbons and substances produced therewith are produced from said land, or Lessee is engaged in drilling or re-working operations on said land.

3. Lessee shall pay royalties to Lessor as follows: (a) one-eighth ($\frac{1}{8}$th) of the Oil produced and saved from said land to be delivered at the wells or to the credit of Lessor into the pipeline to which the well may be connected: Lessee may, at any time or times, purchase any royalty oil, paying the market value in the field on the day it is run to the storage tanks or pipeline: (b) the market value at the well of one-eighth ($\frac{1}{8}$th) of the gas (including casinghead gas or other gaseous substances) produced from the land and sold, provided that on gas sold at the well the royalty shall be one-eighth ($\frac{1}{8}$th) of the amount realized from such sale: (c) one-tenth ($\frac{1}{10}$th) of the amount realized from the sale of any other substances produced from said land with Oil and Gas.

Where there is a gas well or wells on the lands covered by this Lease or acreage unitized therewith, whether it be before or after the Primary Term hereof, and such well or wells are shut-in and there is no other production, drilling operations or other operations being conducted capable of keeping this Lease in force under any of its Provisions, Lessee shall pay as royalty to Lessor (and if it be within the Primary Term hereof such payment shall be in lieu of delay rentals), the sum of one dollar ($1) per year net mineral acre, such payment to be made to the Depository Bank hereinafter named on or before the anniversary date of this Lease next ensuing after the expiration of 90 days from the date such well or wells are shut-in, and thereafter on the anniversary date of this Lease during the period such wells are shut-in, and upon such payment it shall be considered that this Lease is maintained in full force and effect. Lessee may use, free of royalty, oil, gas, and water for all operations hereunder.

4. If drilling operations are not commenced on said land on or before _____, 19__ this Lease shall terminate unless Lessee, on or before that date, shall pay or tender to Lessor, or to Lessor's credit in the _____ Bank at _____ or any successor, the sum of _____ Dollars ($_____) which shall extend for one (1) year the time within which such operation may be commenced. Thereafter, annually, in the same manner and upon the same payment or tender called Rental, this Lease may be continued in force and such operations again deferred for successive periods of one (1) year during the Primary Term: Provided, that if any Oil and Gas shall be produced from or any drilling or re-working operations conducted on said land within ninety (90) days prior to any anniversary of this Lease during the Primary Term, the rental occurring on such anniversary date shall be excused and this Lease shall continue in force as though such rental had been paid. Such operations shall be commenced when the first material is moved in or the first work done. Payments or tenders of rentals may be made by mailing or delivering cash, or Lessee's check or draft to Lessor, or to the Depository Bank on or before the date of payment. If the Depository Bank fails or refuses to accept the rental this Lease shall not terminate, nor Lessee be held in default for failure to pay rental unless Lessee fails to pay such rental for thirty (30) days after Lessor has delivered to Lessee a recordable instrument designating another Depository Bank. Any Bank designated as a Depository shall continue as such and as Lessor's agent regardless of changes in ownership of Lessor's Interest and Lessee may pay tender rental jointly to the credit of all parties having any interests. At the option of Lessee all rental payments may be made to _____ one of the parties named herein as Lessor. If Lessee shall in good faith and with reasonable diligence attempt to pay any rental but fails to pay or incorrectly pays part of the rental, this Lease shall not terminate unless Lessee fails to rectify the error or failure within thirty (30) days after written notice of the failure. Lessee may at any time or times surrender this Lease as to all or any part of the land or as to any stratum or strata, by mailing or tendering to Lessor or to the Depository Bank, or by fil-

ing a release or releases in the County Records, and thereby be relieved of all obligations as to the portion surrendered, after which the rental shall be reduced in the same proportion the acreage is reduced.

5. Lessee may at any time or times pool any part or all of said land and Lease or any stratum or strata, with other lands and Leases, stratum or strata, in the same field so as to constitute a spacing unit to facilitate an orderly or uniform well spacing pattern or to comply with any order, rule or regulation of the State or Federal regulatory or conservation agency having jurisdiction. Such pooling shall be accomplished or terminated by filing of record a Declaration of Pooling, or Declaration of Termination of Pooling, and by mailing or tendering a copy to Lessor, or to the Depository Bank. Drilling or re-working operations upon or production from any part of such spacing unit shall be considered for all purposes of this Lease as operations or productions from this Lease. Lessee shall allocate to this Lease the proportionate share of production which the acreage in this Lease included in any such spacing unit bears to the total acreage in said spacing unit.

6. If at any time or times after the Primary Term or before the expiration of the Primary Term all operations, and if producing, all production shall cease for any cause, this Lease shall not terminate if Lessee commences or resumes any drilling or re-working operations, or production, within ninety (90) days after such cessation; provided that payment of rental as herein provided for shall be resumed if such cessation occurs during the Primary Term, which rental shall be in addition to any royalty paid. Lessee may, in the interest of economy, commingle production from this Lease with production from one or more Leases in the same field provided a method of measurement in accordance with established engineering practices is used to measure the production and to allocate the production to the respective Leases commingled.

7. Lessee shall pay for all damages caused by Lessee's operations to growing crops, buildings, irrigation ditches and fences. When requested by the surface owner, Lessee shall bury pipelines below ordinary plow depth across cultivated lands. No well shall be drilled within two hundred (200) feet of any residence or barn now on said land without the consent of the surface owner. Lessee shall have the right at any time to remove all Lessee's property and fixtures, includ-

ing the right to draw and remove all casing. Lessee shall drill any well which a reasonably prudent operator would drill under the same or similar circumstances to prevent substantial drainage from said land by wells located on adjoining land not owned by Lessor, when such drainage is not compensated by counterdrainage, subject to the continuing right of the Lessee to release all or part of the lands covered hereby as provided for in Paragraph four (4) above. No default of Lessee with respect to any well or part of the land covered hereby shall impair Lessee's rights as to any other well or any other part of the lands covered hereby.

8. The rights of Lessor and Lessee may be assigned in whole or in part. No change in ownership of Lessor's interest shall be binding on Lessee until after Lessee has been given notice consisting of certified copies or recorded instruments or documents necessary to establish a complete chain of title from Lessor. No other type of notice, whether actual or constructive, shall be binding on Lessee, and Lessee may continue to make payments as if no change had occurred. No present or future division of Lessor's ownership as to all or any part of said lands shall enlarge the obligations or diminish the rights of Lessee, and Lessee may disregard any such division. If all or any part of Lessee's interest is assigned, no leasehold owner shall be liable for any act or ommission of any other leasehold owner, and failure by one to pay rental shall not affect the rights of the others; rental is apportionable in proportion to acreage owned by each leasehold owner.

9. Whenever, as a result of any cause reasonably beyond Lessee's control, such as fire, flood, windstorm or other Act of God, decision law, order, rule, or regulation of any local, State or Federal Government or Governmental Agency, or Court; or inability to secure men, material or transportation, and Lessee is thereby prevented from complying with any express or implied obligations of this Lease, Lessee shall not be liable in damages or forfeiture of this Lease, and Lessee's obligations shall be suspended so long as such cause persists, and Lessee shall have ninety (90) days after the cessation of such cause in which to resume performance of this Lease.

10. Lessee may at any time or times unitize all or any part of said land and Lease, or any stratum or strata, with other lands and Leases in the same field so as to constitute a unit or units whenever, in Lessee's judgment, such unitization is required to prevent waste or pro-

APPENDIX

mote and encourage the conservation of Oil and Gas by any cooperative or unit plan of development or operation; or by a cycling,
pressure-maintenance, repressuring or secondary recovery program.
Any such unit formed shall comply with the local, State and Federal
Laws and with the orders, rules, and regulations of State or Federal
regulatory or conservative agency having jurisdiction. The size of
any such unit may be increased by including acreage believed to be
productive, and decreased by excluding acreage believed to be unproductive, or where the owners of which do not join the unit, but
any such change resulting in an increase or decrease of Lessor's royalty shall not be retroactive. Any such unit may be established, enlarged or diminished and in the absence of production from the unit
area, may be abolished and dissolved by filing of record an instrument so declaring, and mailing or tendering to Lessor, or to the Depository Bank, a copy of such instrument. Drilling or re-working
operations upon, or production from any part of such units shall be
considered for all purposes of this Lease as operations or production
from this Lease. Lessee shall allocate to the portion of this Lease
included in any such unit a fractional part of production from such
unit on any one of the following basis': (a) the ratio between the
participating acreage in this Lease in such units and total of all participating acreage in the unit; or, (b) the ratio between the quantity of
recoverable production from the land in this Lease in such unit and
the total of all recoverable production from all such unit; (c) any
basis approved by State or Federal authorities having jurisdiction.
Lessor shall be entitled to the royalties in this Lease on the part of the
unit production so allocated to that part of this Lease included in
such unit and no more.

11. Lessor warrants and agrees to defend the title to said land as
to Lessor's interest therein. The royalties and rental provided for
are determined with respect to the entire mineral estate in Oil and
Gas (including all previously reserved or conveyed non-participating
royalty), and if Lessor owns a lesser interest, the royalty and rental to
be paid Lessor shall be reduced proportionately. Failure of Lessee to
proportionately reduce the rental shall have no relationship to the
royalties to be proportionately reduced if production is secured.
Lessee may purchase or discharge in whole or in part any tax, mortgage or lien upon said land, or redeem the land from any purchaser
at any tax sale or adjudication, and shall be subrogated to such lien

with the right to enforce it, and may reimburse itself from any rentals or royalties accruing under the terms of this Lease.

12. This Lease shall be binding upon all who execute it, whether they are named in the granting clause and whether all parties named in the granting clause execute the Lease or not. All Provisions of this Lease shall inure to the benefit of and be binding upon the heirs, executors, administrators, successors, and assigns of the Lessor and Lessee.

IN WITNESS WHEREOF this instrument is executed on the date first hereinabove set out.

_____   _____

_____   _____

_____   _____

(ACKNOWLEDGMENT)

*

INDEX

References are to Pages

ABANDONMENT OF RIGHTS
Generally, 24–27

ACCOMMODATION DOCTRINE
Generally, 162–165

ACREAGE CONTRIBUTION AGREEMENT
See Support Agreements

***AD COELUM* DOCTRINE**
Generally, 8–9, 365

ADVERSE POSSESSION
    Generally, 70–78
Amount of minerals earned, 77–78
Color of title, 71
Elements of, 71, 74–75
Minerals earned, 78
Paper transactions, 74
Relation back, 73–74
Scope of, 71–74, 77–78
Severance of minerals, 74–76
Tacking, 71
Unity of possession, 72, 77

ALABAMA
Cotenants, right to develop, 82
Duhig rule, 128

AMERICAN ASSOCIATION OF PETROLEUM LANDMEN
Model operating agreements, 351

AMERICAN PETROLEUM INSTITUTE
Model drilling contracts, 355

APPORTIONMENT OF ROYALTIES
Apportionment rule, 141–143, 365
Avoiding problems, 143–147
Entirety clause, 144–146, 372
Non-apportionment rule, 140–143, 377

ARKANSAS
Accommodation doctrine, 164
Life tenant, allocation of bonus, 100
Lease, temporary cessation of production, 245
Market value/proceeds royalty problem, 266–267
Royalty, non-apportionment of, 141

ASSIGNMENT CLAUSE
See Notice of Assignment Clause

ASSIGNMENTS
See Lease Transfers

ASSUMPSIT
See Trespass

BONUS
Generally, 34, 45, 100, 173, 193
Defined, 34, 366
Taxation, 334–335

BOTTOM HOLE AGREEMENT
See Support Agreements

BROWN, EARL
Further exploration covenant, 292

CALIFORNIA
Abandonment, 25, 26–27
Community property, 91
Cotenants, right to develop, 82
Lease,
    "Or" drilling-delay rental clause, 185, 186–187
    Profit a prendre, 153
    Size of royalty, 258
Non-ownership theory, 23
Pooling, as a cross-conveyance, 231
"Subject to Problem", 136

CANADIAN ASSOCIATION OF PETROLEUM LANDMEN
Model operating agreement, 351

CAPTURE, RULE OF
See Rule of Capture

CARRIED INTEREST
    Generally, 21, 41–43
Defined, 366

CASINGHEAD GAS
Covered by lease, 155
Defined, 367

CESSATION OF PRODUCTION CLAUSE
Defined, 245–246, 367
Example, 246
Interpretation, 246–248
Permanent cessation, 244–245
Temporary Cessation of Production Doctrine, 244–245

COLORADO
Acknowledgement, effect of defective, 67
Duhig rule, 128
Mineral interests, lost or abandoned, 89
Mineral/royalty distinction, 120–121
Non-ownership theory, 23

COLORADO—Continued
Royalty, non-apportionment of, 141

COMMENCEMENT OF DRILLING
See Drilling-Delay Rental Clause

COMMON, TENANCY IN
See Concurrent Owners

CONCURRENT OWNERS
     See also, Trespass
  Generally, 79–90, 152
Development by,
    Critique of majority rule, 83–84
    Majority rule, 81–82
    Minority rule, 81
    Ouster, 83
Dormant minerals acts, 89
Forced Pooling, 85–86, 87
Joint tenancy, 79–80, 83
Lost mineral interests, 87–90, 154–155
Marketable record title acts, 89
Partition, 86–87
Prescription, 88, 381
Tenancy by the entirety, 80, 83
Tenancy in common, 79, 80, 82, 83
Waste, 83

CONSERVATION LAWS
  Generally, 14–22, 85–86, 167
Compulsory pooling, 20–21
Exception tract statutes, 21
Gas/oil ratio, 19
Production allowables, 19
Prorationing, 19
Purpose, 17
Water/oil ratio, 19
Well-spacing rules, 19

CONSIDERATION
See Conveyances, Execution

CONSTRUCTION OF INSTRUMENTS
See Judicial Construction

CONTINUOUS OPERATIONS CLAUSE
See Operations Clause

CONVERSION
See Trespass

CONVEYANCES
Generally, 52–69, 150, 152, 154, 156
Adverse possession, 70–77
Covenants of title, 53–57
Deed polls, 64
Deeds, 53–55, 64, 65, 67, 69
    Limited warranty, 54
    Quitclaim, 54–55
    Warranty, 53–54
Description,
    Legal validity, 58–59
    Marketability, 59, 62
    Methods of, 58–62
Designation of grantor and grantee, 62–63
Elements of, 52
Execution,
        Generally, 63–69
    Acceptance, 67–68
    Acknowledgement, 65
    Attestation, 65
    Consideration, 64–65
    Defects, effect of, 66
    Delivery, 67
    Signature, 63–64
Inheritance, 69
Judicial transfers, 69–70
Leases as, 53, 56–57, 64, 65, 67, 152

CONVEYANCES—Continued
Recording statutes, 66, 69
Statute of Frauds, 53, 63, 196
Writing requirement, 53
Warranties of title, 53–57

CORPOREAL/INCORPOREAL DISTINCTION
Abandonment, 24–27
Forms of Action, 27–28

CORRELATIVE RIGHTS DOCTRINE
Generally, 13–14, 85
Defined, 368
Limitations, 17

COTENANCY
See Concurrent Owners

COVENANTS OF TITLE
Implied in deeds, 54
Implied in leases, 249
Lease clause, 249–251
Warranty deed, 53–54

COVER ALL CLAUSE
See Granting Clause

CREDITORS/DEBTORS
Conveyances by, 92–94

DEBTORS/CREDITORS
Conveyances by, 92–94

DEEDS
See Conveyances

DELAY RENTAL CLAUSE
See Drilling-Delay Rental Clause

DELAY RENTALS
See Drilling-Delay Rental Clause

DEPLETION
See Taxation

DESCRIPTION
See Conveyances

DIVISION ORDER
Generally, 362
Amendment of lease by, 272–274
Defined, 369
Ratification of gas contract by, 272–274

DOUBLE FRACTION PROBLEM
Generally, 123–125
Defined, 123–124, 370

DRAINAGE, COVENANT TO PROTECT AGAINST
See Implied Covenants, Protection Covenant

DRILLING CONTRACT
Generally, 353–354, 370
Daywork contract, 354
Footage contract, 354
Turnkey contract, 354

DRILLING–DELAY RENTAL CLAUSE
Generally, 181–208
Delay rental, defined, 369
Drilling option,
      Generally, 189–192
Commencement, meaning of, 189–190, 191, 192, 194, 216
Completion, 190
Due diligence required, 191–192
Good faith, 191, 192
Sham operations, 191
Examples of, 184, 185

DRILLING–DELAY RENTAL CLAUSE—Continued
Judicial protection of lessee,
    Estoppel, 195–198
    Revivor, 195–196
    Waiver, 195–198
No term lease, 183
"Or" lease,
        Generally, 185–189
    Defined, 185, 379
    Forfeiture clause, 186–189, 206
    Remedy for failure to pay, 185–189
Paid-up lease, 206–207
Payment of delay rentals,
        Generally, 193–208, 249
    Amount, 193
    Clerical mistake, 193
    Failure to pay, 193–195, 206, 325
    Good faith mistake, 194
    Strict standard, 193–194
Protective lease clauses,
    Notice of assignment clauses, 200–203, 253–254
    Notice of nonpayment clauses, 203–205
    "Or" leases, 205–206
"Unless" lease,
        Generally, 183–185
    Automatic termination, 184–185
    Defined, 184, 387
    Equitable protections, 195–198
    Lease protective clauses, 198–208
    Special limitation to primary term, 184

DRILLING OPERATIONS CLAUSE
See Operations Clause

DRY HOLE AGREEMENT
See Support Agreements

DRY HOLE CLAUSE
Generally, 210–215
Defined, 371
Dry hole definition, 212
Example of, 210–211
When payment due, 212–215

DUHIG RULE
Generally, 126–132
Application to leases, 131–132
Avoiding the problem, 132–133
Defined, 128, 371
Departures from, 129–130
Warranty is key, 128

EASEMENTS
See Granting Clause, Surface uses granted

EJUSDEM GENERIS RULE
See Judicial Construction

ELLIS, WILLIS
Duhig rule, 130

ENTIRETIES, TENANCY BY
See Concurrent Owners

EQUIPMENT REMOVAL CLAUSE
Generally, 253–254

EXCEPTION TRACTS
See Conservation Laws

EXECUTIVE RIGHTS
Conveyances by, 95–96
Defined, 95

EXPLORATION COVENANT
See Implied Covenants

FARMOUT AGREEMENTS
Generally, 348–350
Defined, 372
Purpose of, 348
Substantive issues in, 349
Taxation of, 338–339

FEDERAL INCOME TAXATION
See Taxation

FEE INTEREST
Generally, 31–32, 373

FIDUCIARIES/BENEFICIARIES
Conveyances by, 94–95

FLORIDA
Cotenants, right to develop, 82

FORCE MAJEURE CLAUSE
Generally, 236–238
Defined, 236, 373
Example of, 236–237

FORCED POOLING
See Concurrent Owners; Conservation Laws, Compulsory
Pooling

FOUR CORNERS RULE
See Judicial Construction

FRACTIONAL INTERESTS
See Double Fraction Problem; *Duhig* Rule; Mineral
Acres/Royalty Acres

FREESTONE RIDER
See Pooling/Unitization Problems

GAS BALANCING AGREEMENTS
Generally, 361–362
Defined, 374

GAS CONTRACTS
See also, Market Value/Proceeds Royalty Problem
Generally, 356–362, 375
Commitment, 360
Conditions of delivery, 361
Price,
Generally, 357–359
Area rate clause, 357, 366
Buyer protection clauses, 358
Deregulation clause, 358, 369
Ferc-out clause, 359, 373
Fixed escalation clause, 357
Index-escalation clause, 358
Market-out clause, 359, 376
Most favored nations clauses, 357
Price redetermination clause, 358, 381
Reservations, 360
Substantive issues, 356–361
Take or pay, 359

GEORGIA
Cotenants, right to develop, 82

GRANTING CLAUSE
Generally, 26, 36, 52–68, 154–171
Cover all clause, 156–157, 368
Defined, 375
Description, 58–62
Double fractions, 123–126
Mineral/royalty distinction, 117–123
Mother Hubbard clause, 156–159
Overconveyance, 126–133
Size of interest, 154–155
Surface uses granted,
Generally, 159–171

GRANTING CLAUSE—Continued
Surface uses granted—Continued
  Accommodation doctrine, 162–165
  Benefit to minerals, 165–167
  Implied easement, 159–160
  Lease terms, 169–170
  Legislative provisions, 170–171
  Reasonable use, 159–162, 170–171
Words of grant, 57–58

HABENDUM CLAUSE
See Term Clause

ILLINOIS
Cotenants, right to develop, 81
Pooling, as a cross-conveyance, 231
Royalty, non-apportionment of, 141

IMPLIED COVENANTS
Cooperation principle, 277
Covenant to test, 282
Further exploration covenant,
    Generally, 291–298
  Defined, 374
  Distinguished from reasonable development covenant,
    292–295
  Enforcement, 296–297
  Probability of profit, 292–295
  Proof of breach, 295–296
  Remedies for breach, 297–298
Implied in fact, 276–277
Implied in law, 276–277
Marketing covenant,
    Generally, 307–313
  Defined, 307, 376
  Lease provisions affecting, 313
  Proof of breach, 309–310, 312–313
  Reasonable price, 309
  Reasonable time, 308

IMPLIED COVENANTS—Continued
Marketing covenant—Continued
    Remedies for breach, 310–311
Operating covenant, 314–315
Protection covenant,
        Generally, 230, 298–306, 313
      Common lessee problem, 301–303
      Conservation laws affecting, 298
      Defined, 298, 382
      Lease provisions affecting, 304–306
      Probability of profit, 301
      Proof of breach, 300–306
      Remedies for breach, 306–307
Reasonable development covenant,
        Generally, 283–291
      Defined, 283, 383
      Imprudent operator, 284–286
      Lease provisions affecting, 287–289
      Probability of profit, 284
      Proof of breach, 283–289
      Remedies for breach, 289–291
Reasonable prudent operator standard, 279–281, 382

IN PAYING QUANTITIES
See Term Clause, Production

INCOME TAXATION
See Taxation

INDIANA
Dormant minerals act, 89
Royalty, non-apportionment of, 141

INTANGIBLE DRILLING COSTS
See Taxation

**INTERNATIONAL ASSOCIATION OF DRILLING CONTRAC-
      TORS**
Model drilling contracts, 355

[*423*]

JUDICIAL ASCERTAINMENT CLAUSE

Generally, 256–257

JUDICIAL CONSTRUCTION

See also, Double Fraction Problem; *Duhig* Rule; Mineral/Interest; Minerals, Meaning of; Mineral Acre; Royalty Acre

Generally, 105–107

Construction aids, 106–107, 109–111

Ejusdem Generis, 107

Extrinsic evidence, 107

Four corners rule, 106, 109

Intent of parties controls, 105

Interpretation against preparer, 106, 107

Steps in interpretation, 105–106, 108

KANSAS

Cotenants, right to develop, 82

Lease,

As a license, 153

Meaning of "production," 174–175

Substances covered, 155

Market value/proceeds royalty problem, 265

Mineral deed, recorded or void, 69

Ownership in place theory, 23

Partition, 87

Pooling, rule against perpetuities, 234

Royalty, non-apportionment of, 141

Term interest, 103

KENTUCKY

Cotenants, right to develop, 82

Lease,

Commencement of operations, 215

Meaning of "production", 175

Royalty, non-apportionment of, 141

KUNTZ, EUGENE
Accommodation doctrine, 163
Dry hole definition, 212
"Minerals," meaning of, 115–116
Shut-in royalty clause, 244

LANDOWNERS ROYALTY
See Royalty Interest

LEASE TRANSFERS
Assignment, 317–318, 320, 321, 322, 323
Assignment/sublease distinction, 320
Covenants running with land, 319
Delay rentals, divisibility of, 325–326
Exculpatory language, 318–319
Implied covenants, divisibility of, 326–327
Implied covenants of title, 323–324
Indemnification of lessee by assignee, 318
Lease obligations, divisibility of, 318, 326
Lessee's right to transfer, 317–318
Non-operating interests, protection of, 321–323
Partial assignee's failure to perform, 324–326
Washout, protection of non-operating interests, 322–323

LEASES
See also, Drilling-Delay Rental Clause; Granting
Clause; Market Value/Proceeds Royalty
Problem; Royalty Clause; Term Clause
Generally, 25, 53, 56–57, 64, 65, 67, 150–327
Essential provisions, 154
Indefinite duration, 151
Interest created, 153
Nature of, 25, 152–154
Option to develop, 150–151
Purpose of, 150, 276

LESSER INTEREST CLAUSE
Generally, 251–252

LESSER INTEREST CLAUSE—Continued
Defined, 376

LIFE TENANT/REMAINDERMAN
Corpus/income distinction, 100
Division of lease proceeds, 99–100
Leasing from, 98–100
Open mine doctrine, 101
Power to convey, 97–100

LIMITATION TITLE TO MINERALS
See Adverse Possession

LOST MINERAL INTERESTS
See Concurrent Owners

LOUISIANA
Adverse possession, 74
Community property, 91
Co-tenants, right to develop, 81
Duhig rule, 128
Executive right, 95–96
Lease,
    Assignment/sublease distinctions, 320
    Cancellation for failure to pay royalty, 275
    Contract, not conveyance, 152
    Divisibility of obligations, 326
    Nature of royalty, 260
Market value/proceeds royalty problem, 266–267
Mineral royalty, 39
Mineral servitude, 35, 103
Naked owner, 98
Non-ownership theory, 23
Open mine doctrine, 102
Pooling,
    Cross-conveyance, 231
    Pugh clause, 235–236
    Ratification by non-executive, 233
    Royalty, lessee's right to pool, 221

LOUISIANA—Continued
Prescription, 35, 88, 103
Pugh clause, 235–236
Reasonable development covenant,
      Generally, 285–286
      Cancellation remedy, 289
      Necessity of notice, 287
Royalty, non-apportionment of, 141
Usufruct, 98

MARKET VALUE/PROCEEDS ROYALTY PROBLEM
      Generally, 263–274
Basis of dispute, 264–265
Current market value theory, 265–266
Proceeds theory, 266–267
Solutions, 269–274

MARKETING COVENANT
See Implied Covenants

MARRIED PERSONS
      Generally, 90–92
Community property, 91–92
Conveyances by, 92
Curtesy, 90
Dower, 90, 92
Homestead, 91, 92

MARTIN, PATRICK H., JR.
Implied covenants, 316

MAXWELL, RICHARD
Mineral/royalty distinction, 121

MEYERS, CHARLES
      See also, Williams, Howard, and Meyers, Charles
Further exploration covenant, 292, 294

MICHIGAN
"Commencement" of operations, 191

MINERAL ACRE
 Generally, 133–134
Defined, 376

MINERAL DEED
See Conveyances; Deeds

MINERAL INTEREST
 Generally, 32–35, 72, 79, 81, 118, 123
Defined, 33, 377
Easement over surface, 33, 37, 159–160
Elements of, 33–35, 116
Executive rights, 34, 116
Mineral/royalty distinction, 117–123
Reservations of, 35
Substances defined as, 108–117
Surface use rights, 32–33, 116, 159–171

MINERALS, MEANING OF
 Generally, 108–117
Casinghead gas, 155
Community knowledge test, 110
Exceptional characteristics test, 110–111
Helium, 109, 155
Lease, when used in, 155–156
Lignite, 112
Manner of enjoyment theory, 115–117, 163
Oil and gas, 109
Rule of practical construction, 110–111
Surface destruction test, 111–115
Texas approach, 111–115

MISSISSIPPI
Duhig rule, 128
Pooling, as a cross-conveyance, 231
Royalty, apportionment of, 141

MISSISSIPPI—Continued
Rule of capture, 12

MISSOURI
Cotenants, right to develop, 82

MONTANA
Cotenants, right to develop, 82
Lease,
    Commencement of drilling/operations, 190
    Meaning of "production," 175
    Profit a prendre, 153
Market value/proceeds royalty problem, 266
Pooling, as a cross-conveyance, 231

MOTHER HUBBARD CLAUSE
See Granting Clause

NEBRASKA
Royalty, non-apportionment of, 141
Rule of capture, 12

NET PROFITS INTEREST
    Generally, 40–41
Defined, 377

NEVADA
Lease, effect of delay rental provision, 182
Rule of capture, 112

NEW MEXICO
Community Property, 91
Duhig rule, 128
Mineral/royalty distinction, 119
Non-ownership theory, 23
Royalty, non-apportionment of, 141

NO INCREASE OF BURDEN CLAUSE
Generally, 254–255

INDEX
References are to Pages

NO TERM LEASE
See Drilling-Delay Rental Clause

NON–APPORTIONMENT RULE
See Apportionment of Royalties

NON–EXECUTIVE RIGHTS
Conveyances by, 95–96
Defined, 95

NON–OWNERSHIP THEORY
Generally, 22–23, 25, 28–30
Defined, 378
Significance,
　　Abandonment, 24–27
　　Forms of action, 27–28
　　Practical, 29–30
　　Real property/personal property, 28–29

NON–PARTICIPATING INTEREST
See Non-Executive Rights; Royalty Interest

NON–PARTICIPATING ROYALTY
See Royalty Interest

NORTH DAKOTA
Accommodation doctrine, 164
Cotenants, right to develop, 82
Duhig rule, rejection, 129
Lease, cancellation for failure to pay royalty, 275
Market value/proceeds royalty problem, 266

NOTICE BEFORE FORFEITURE CLAUSE
Generally, 256–257

NOTICE OF ASSIGNMENT CLAUSE
Generally, 200–203, 254

NOTICE OF NONPAYMENT CLAUSE
Generally, 203–205

OFFSET WELL COVENANT
See Implied Covenants; Protection Covenant

OHIO
Royalty, non-apportionment of, 141

OKLAHOMA
Cotenants, right to develop, 82
Duhig rule, 128
Forced pooling, 86
Further exploration covenant, rejection, 293–294
Lease,
      Commencement of operations, 215
      Meaning of "production", 175, 179–180
      Notice of non-payment clause, 203
      Profit a prendre, 153
      Shut-in royalty, failure to pay, 244
      Substances covered, 155
      Temporary cessation of production, 246–248
Life tenant, allocation of bonus, 100
Market value/proceeds royalty problem, 266–267
Mineral interests, lost or abandoned, 90
Non-ownership theory, 23
Open mine doctrine, 101
Partition, 87
Pooling as a cross-conveyance, 231
Reasonable development covenant, proof burden, 283
Royalty, non-apportionment of, 141
Rule of capture, 11
Term interest, 103

ONTARIO
Royalty, apportionment of, 141

OPEN MINE DOCTRINE
See Life Tenant/Remainderman

OPERATING AGREEMENT
Generally, 350–353
Defined, 379
Model agreements, 351
Substantive issues, 351–353

OPERATIONS CLAUSE
Generally, 215–220
Continuous operations clause, 216–217
Defined, 379
Delay between completion and production, 219–220
Purpose, 215–216
Well completion clause, 216–217

"OR" LEASE
See Drilling-Delay Rental Clause

OVERCONVEYANCE PROBLEM
See *Duhig* Rule

OVERRIDING ROYALTY
See Royalty Interest

OWNERSHIP IN PLACE THEORY
Generally, 22–23, 25, 28–30
Defined, 380
Significance,
Abandonment, 24–27
Forms of action, 27–28
Practical, 29–30
Real property/personal property, 28–29

PAID–UP LEASE
See Drilling-Delay Rental Clause

PARTITION
See Concurrent Owners

PAYING QUANTITIES
See Term Clause

PENNSYLVANIA
Cotenants, right to develop, 82
Lease, "Or" drilling-delay rental clause, 186
"Minerals," oil and gas as, 109
Royalty, apportionment of, 141

PERCENTAGE DEPLETION
See Taxation

PERMEABILITY
Generally, 2, 151

PERSONAL PROPERTY
See Real Property/Personal Property Distinction

PETROLEUM
Casing, 5, 367
Defined, 1
Drilling rigs, 3–5
Formations, 1–2
Reservoirs, 2–3

POOLING CLAUSE
Generally, 220–236
Community lease, 222
Defined, 220, 381
Effect of, 224
Example of, 223–224
Lack of, effect, 221

POOLING/UNITIZATION PROBLEMS
Cross-conveyance theory, 231–232
Duty to exercise, 230–231
Freestone Rider, 234–236, 374

POOLING/UNITIZATION PROBLEMS—Continued
Good faith exercise, 228–229
Non-executive's right to ratify, 232–233
Pugh clause, 234–236, 382
Rule against perpetuities, 234
Timely unit designation, 229–230

POROSITY
Generally, 2, 151

PRESCRIPTION
See Louisiana

PRIMARY TERM
See Term Clause

PRODUCING OIL AND GAS
Casing, 5, 367
Christmas tree, 6
Horsehead, 6, 375
Perforating sonde, 5–6
Pumping unit, 6
Recovery techniques, 7

PRODUCTION, MEANING OF
See Term Clause

PRODUCTION PAYMENT
Generally, 40
Defined, 382
Taxation of, 338

PROFIT A PRENDRE
Generally, 22–23, 25, 56, 153

PROPORTIONATE REDUCTION CLAUSE
Generally, 251–252
Defined, 382

PRORATIONING
See Conservation Laws

PROTECTION COVENANT
See Implied Covenants

PRUDENT OPERATOR STANDARD
See Implied Covenants

PUGH CLAUSE
See Pooling/Unitization Problems

QUALIFIED OWNERSHIP THEORY
See Non-Ownership Theory

REAL PROPERTY/PERSONAL PROPERTY DISTINCTION
Generally, 28–29

REASONABLE DEVELOPMENT COVENANT
See Implied Covenants

ROYALTY ACRE
Generally, 134

ROYALTY CLAUSE
    See also, Market Value/Proceeds Royalty Problem
Amount of royalty, 258, 259
Cancellation for failure to pay, 275
Costs of production/subsequent to production, 262
Deductions from royalty, 261–263
Distinction between oil royalty and gas royalty, 260
Example of, 259
Failure to pay, 274–275
Function of, 258
Market value/proceeds problem, 263–274
Nature of interest created, 260–261

ROYALTY INTEREST
    Generally, 26, 29, 37–40, 56, 117–118, 132

INDEX
References are to Pages

ROYALTY INTEREST—Continued
Apportionment, 138–143
Characteristics of, 39–40, 117–118, 383
Deed forms, 122
Defined, 37–38, 117–118, 383
Interpretive problems, 118–119
Landowner's royalty, 38, 258–275, 375
Mineral royalty, Louisiana, 39
Mineral/royalty distinction, 117–123
Non-participating royalty, 38, 58, 378
Overriding royalty, 38, 176, 379
Perpetual royalty, 39
Surface use rights, 122
Term royalty, 38, 386

RULE OF CAPTURE
Generally, 9–10, 44, 141, 224
Defined, 384
Limitations,
Conservation laws, 14–22
Correlative rights doctrine, 13–14
Drainage by enhanced recovery, 11–13
Escaped hydrocarbons, 10–11
Non-apportionment rule, basis of, 140

SECONDARY TERM
See Term Clause

SEDIMENTARY ROCK
Generally, 2

SEISMIC STUDIES
Generally, 2, 342

SEPARATE OWNERSHIP CLAUSE
Generally, 255

SHUT–IN ROYALTY CLAUSE
Generally, 238–244
Defined, 239, 384
Duration of payments, 242
Effect, 240
Example of, 239–240
Failure to pay, effect, 242–244
Meaning of "shut-in", 241–242
Scope of clause, 240

SLANDER OF TITLE
See Trespass

SMALL TRACT PROBLEM
See Conservation Laws, Exception tract statutes

STATUTE OF FRAUDS
See Conveyances

"SUBJECT TO" PROBLEM
Generally, 135–138
Avoiding, 137–138
Hoffman clause, 138
Lease benefits, 135
Two grants doctrine, 136–138

SUBROGATION CLAUSE
Generally, 252–253
Defined, 385

SUCCESSIVE INTERESTS
See Life Tenant/Remainderman

SUPPORT AGREEMENTS
Generally, 347–348, 385
Acreage contribution agreement, 348, 365
Bottom hole agreement, 348, 366
Dry hole agreement, 347–348, 371
Taxation, 339

SURFACE DAMAGE CLAUSE
See Granting Clause, Surface uses granted

SURFACE DAMAGES ACTS
Generally, 170–171

SURFACE INTEREST
Generally, 36–37, 56, 72
Defined, 385   .
Servient to mineral interest, 159–171
Substances owned, 108–117

SURFACE USE
See Granting Clause

SURRENDER CLAUSE
Generally, 255–256
Defined, 385

TAKE OR PAY
See Gas Contracts

TAXATION
Generally, 328–345
Basic principles, 328–334
Bonus payments, 334
Complete payout concept, 331
Concurrent development, 341
Corporate form, 339
Delay rentals, 335
Depletion,
    Cost depletion, 331–332, 368
    Economic interest requirement, 333
    Independent producers limitation, 333–334
    Percentage depletion, 332, 380
    Royalty owners limitation, 333–334
Equipment costs, 343
Geological and geophysical costs, 343–344
Intangible drilling costs, 329–330, 375

TAXATION—Continued
Partnerships, 340
Production payment, 337–338, 382
Property concept, 329
Revenue Ruling 77–176, p. 338
Royalty payments, 335
Sale/sublease distinction, 335–337
Sharing arrangement, 338
Subchapter S corporation, 340
Windfall Profit Tax, 344–345

TEMPORARY CESSATION OF PRODUCTION DOCTRINE
See Cessation of Production Clause

TENANCY BY THE ENTIRETY
See Concurrent Owners

TENANCY IN COMMON
See Concurrent Owners

TENNESSEE
Lease,
        Meaning of "production", 175
        Primary term limit, 88–89

TERM CLAUSE
        Generally, 174–181
Defined, 172, 386
Example of, 172
Primary term,
        Generally, 172–173, 182, 211
        Length of, 172–173
        Purpose, 173
Production, in paying quantities,
        Equitable considerations, 181
        Expenses considered, 178–180
        Meaning of, 176–177
        Purpose, 176
        Revenues considered, 177–178

TERM CLAUSE—Continued
Production, in paying quantities—Continued
    Termination of, 180–181
    Time period to measure, 180–181
Production, meaning of,
    Actual production required, 175, 215
    Discovery is enough, 175
    In paying quantities, 175–181
Secondary term,
      Generally, 173–181
    Indefinite length, 173, 176
    Purpose, 173, 175, 176

TERM INTERESTS
Conveyances by, 102–104

TEXAS
Accommodation doctrine, 162
Community property, 91
Conservation law, exception tracts, 21
Cotenants, right to develop, 82
Description systems, 60
Duhig rule, 128, 131
Executive right, pooling, 96
Forced pooling, 86
Further exploration covenant, 288–289
Lease,
    Fee simple determinable, 153
    Guardian as grantor, 94
    Implied warranty, 249
    Meaning of "production", 176–177
    Mother Hubbard clause, 159
    Notice of nonpayment clause, 204
Market value/proceeds royalty problem, 265–274
Marketing covenant, proof, 309–311
"Minerals," meaning of, 111–117
Open mine doctrine, 101
Ownership in place theory, 23

TEXAS—Continued
Pooling,
 Community lease, effect of, 222
 Cross-conveyance, 231
 Freestone rider, 235
 Good faith requirement, 228
 Ratification by non-executive, 232
Protection covenant, common lessee problem, 298–300
Reasonable development covenant, limitations, 288–289
Royalty, non-apportionment of, 141
Rule of capture, 12, 13–14
"Subject to" problem, 136
Term interest, extension, 103

TOP LEASE
 Generally, 147–149
Bottom lease, 147
Defined, 386
Obstruction, 149
Rule against perpetuities, 148

TRESPASS
 Generally, 44–51
Bad faith, 49–50
Benefit test, 50
Good faith, 50–51
Remedies for,
 Assumpsit, 48–49, 51
 Conversion, 49, 51
 Damage to lease value, 44–46, 51
 Ejectment, 49, 51
 Slander of title, 47–48, 51

TWO GRANTS DOCTRINE
See "Subject To" Problem

UNITIZATION CLAUSE
 Generally, 224–227

UNITIZATION CLAUSE—Continued
Defined, 220–221, 387
Example of, 225–227

"UNLESS" LEASE
See Drilling-Delay Rental Clause

USUFRUCT
See Louisiana

UTAH
Accommodation doctrine, 164
Duhig rule, 129–130
Pooling, rule against perpetuities, 234

WARRANTY
Clause in lease,
      Generally, 249–251
    After-acquired title, 250
    Covenant of warranty only, 250
    Defined, 387
    Disclaimer, 251
Title covenants, 55–56
Warranty deed, 53–54

WASH OUT TRANSACTIONS
See Lease Transfers

WELL COMPLETION CLAUSE
See Operations Clause

WELL-SPACING RULES
See Conservation Laws; Pooling Clause

WEST VIRGINIA
Cotenants, right to develop, 81
Lease, meaning of "production", 175
Pooling, as a cross-conveyance, 231
Reasonable development covenant, damages, 291
Royalty, non-apportionment of, 141

WILLIAMS, HOWARD, AND MEYERS, CHARLES
Adverse possession, lease as mineral severance, 76
Implied covenants, sources of, 277
Operating covenant, 314
Protection covenant, damages, 307
Reasonable development covenant, damages, 291
Royalty acres, 134

WORKING INTEREST
Generally, 36
Defined, 388

WYOMING
Duhig rule, 128
Non-ownership theory, 23
Lease,
Meaning of "production", 175
Profit a prendre, 153

†